Cross-Cultural Skills for Deployed Air Force Personnel

Defining Cross-Cultural Performance

Chaitra M. Hardison, Carra S. Sims, Farhana Ali, Andres Villamizar, Ben Mundell, Paul Howe

Prepared for the United States Air Force
Approved for public release; distribution unlimited

RAND PROJECT AIR FORCE

The research described in this report was sponsored by the United States Air Force under Contract FA7014-06-C-0001. Further information may be obtained from the Strategic Planning Division, Directorate of Plans, Hq USAF.

Library of Congress Cataloging-in-Publication Data is available for this publication.

ISBN 978-0-8330-4606-2

Published 2009 by the RAND Corporation
1776 Main Street, P.O. Box 2138, Santa Monica, CA 90407-2138
1200 South Hayes Street, Arlington, VA 22202-5050
4570 Fifth Avenue, Suite 600, Pittsburgh, PA 15213-2665
RAND URL: http://www.rand.org/
To order RAND documents or to obtain additional information, contact
Distribution Services: Telephone: (310) 451-7002;
Fax: (310) 451-6915; Email: order@rand.org

Preface

Air Force senior leadership recognizes that cross-cultural competence is critical in preparing airmen to accomplish a variety of missions in foreign environments. This recognition has led to strong interest in ensuring that the Air Force provides the cross-cultural skills its airmen need through training and education. RAND was asked to provide a definition of cross-cultural performance as the foundation for the design of the Air Force's cross-cultural training and education. The resulting definition is a first step toward clarifying what airmen should be able to do to be considered cross-culturally competent.

The research reported here was sponsored by the Air Force Directorate of Airman Development (AF/A1D) and conducted within the Manpower, Personnel, and Training Program of RAND Project AIR FORCE as part of a fiscal year 2008 study titled "Building Better Airmen Through Enhanced Force Development Policies and Processes." This monograph should be of interest to those involved in or interested in providing culture and language training for international assignments: Air Force leadership and staff, the broader defense community, government agencies involved in international assignments, and any public or private organizations operating in an international environment.

RAND Project AIR FORCE

RAND Project AIR FORCE (PAF), a division of the RAND Corporation, is the U.S. Air Force's federally funded research and development center for studies and analysis. PAF provides the Air Force with independent analyses of policy alternatives affecting the development, employment, combat readiness, and support of current and future aerospace forces. Research is performed in four programs: Force Modernization and Employment; Manpower, Personnel, and Training; Resource Management; and Strategy and Doctrine. The research reported here was prepared under contract FA7014-06-C-0001.

Additional information about PAF is available at
http://www.rand.org/paf/

Contents

Figures

Tables

Summary

Air Force leadership recognizes that the cross-cultural performance of Air Force members now plays a greater role in mission success than ever before. AF/A1D therefore asked RAND to assist in developing a comprehensive program for preparing members of the Air Force in cross-cultural skills.

To better understand the behaviors that the Air Force hoped to develop and improve through this program, we reviewed existing literature on cross-cultural job performance and discussed cross-cultural training needs with various Air Force personnel. We found that there is no clearly established description of the behaviors required to perform in a foreign country and that the opinions of Air Force personnel differ about what should be included in the training for such performance. Given these differences of opinion and the fact that no taxonomy covering all relevant aspects of cross-cultural performance had yet been established, RAND was given the goal of creating the taxonomy.

To accomplish this goal, we began with three main questions:

1. What is cross-cultural performance, or behavior?
2. Which cross-cultural behaviors do Air Force members, or airmen, identify as important to their deployed jobs?
3. Do all airmen, regardless of job requirements, need the same type and/or amount of cross-cultural training?

We then developed a framework of 14 cross-cultural behavior categories that could be used to address these three main questions.

Nine of the 14 categories encompass behaviors that are considered *enabling*—i.e., they help facilitate a variety of day-to-day activities and are likely to be needed in a variety of jobs. These categories are foreign language skills; verbal and nonverbal communication; applying appropriate social etiquette; managing stress in an unfamiliar cultural setting; changing behavior to fit cultural context; gathering and interpreting observed information; applying regional knowledge; self-initiated learning; respecting cultural differences. The other five categories encompass behaviors that are considered *goal oriented*—i.e., they are associated with specific mission-related activities and are likely to be needed only by individuals working in certain Air Force Specialty Codes (AFSCs) or for certain job tasks. These categories are establishing authority; influencing others; negotiating with others; establishing credibility, trust, and respect; resolving conflict.

The importance of the 14 behavior categories for deployed performance was evaluated by surveying approximately 21,000 previously deployed airmen. The survey respondents generally rated enabling behaviors as more important than goal-oriented behaviors; they also rated skill in the language of the country of deployment as less important than half of the other categories. (See pp. 18–20.) We also examined the importance ratings within each AFSC and the differences across AFSCs. For example, Pilots tended to view the 14 behavior categories as unimportant to their deployed jobs, whereas Special Investigations, International Affairs/Foreign Area Office, and Security Forces personnel tended to have the opposite view. (See pp. 20–23.)

Overall, the results show that at least some airmen view each category of cross-cultural behavior as important (see pp. 18–27), suggesting that the Air Force should include all 14 in a comprehensive training program. Based on the survey findings, we recommend three graduated levels of training in each of the categories. All airmen would receive a low level of training throughout their career and just prior to deployment. Airmen in AFSCs that indicated the 14 categories were, on average, moderately important would also receive medium-level, in-depth training. And airmen in AFSCs that rated certain categories highly important would also receive high-level—specialized, expert-level—training. (See pp. 20–23.)

Other suggestions for next steps in designing a comprehensive cross-cultural training program include evaluating existing Air Force cross-cultural training, designing new curricula to address the 14 behavior categories where necessary, developing assessment tools for measuring skills in the 14 categories, setting standards for cross-cultural performance expectations, and tracking the skills and training received. (See pp. 41–47.)

Acknowledgments

A number of Air Force personnel have contributed to this study. We specifically thank our former and current project sponsors, Brig Gen Robert Allardice (AF/A1D) and Mr. Joseph McDade (AF/A1D), as well as Ms. Gwendolyn DeFilippi (AF/A1DD) and various staff members, including Capt Bryan Pickett (AF/A1DD), Capt Greg Duffy (AF/A1DLL), and LtCol Charlie Underhill (AF/A1DLL), for their guidance and support in sponsoring this study.

We also thank the personnel at the Air Force Manpower Agency, including Mr. Lou Datko, Ms. Brenda Gainey, and Ms. Ashley Brown, for providing us with sampling information and for tirelessly sending out emails inviting participants to complete our survey.

In addition, we would like to thank Mike Thirtle for serving as the project lead at the initiation of this project, laying the groundwork for this study, and providing guidance to the team throughout the remainder of the project. We also acknowledge John Crown, for helping with the Air Force personnel data used for selecting the survey sample, and Larry Hanser and Al Robbert, for their guidance and assistance throughout the study. Col Cathy Chin helped us in the initial stages of the open-ended comments analysis. Julie Ann Tajiri, Emily Taylor, Taylor Forry, and Samantha Abernethy provided tireless support. Cord Thomas and Rita Singer assisted with the technical aspects of fielding the survey. Finally, Mike Neumann contributed to the editing process.

Lastly, but perhaps most importantly, we thank the many airmen who participated in the survey, as well as the faculty members, instructors, career field managers, and Air Force personnel who participated in the informal meetings, interviews, and focus groups leading to the survey's development.

Abbreviations

AF/A1D	Air Force Directorate of Airman Development
AFSC	Air Force Specialty Code
DLI	Defense Language Institute
DoD	Department of Defense
FAO	Foreign Area Officer
OSI	Office of Special Investigations
PAF	Project Air Force
PME	professional military education
ROE	rules of engagement
SOP	standard operating procedure
TCN	third-country national

CHAPTER ONE

Introduction

In today's military climate, cross-cultural performance plays a greater role in mission success than ever before. Both military and civilian leadership have clearly acknowledged the need for further development of cross-cultural performance, including such specific skills, or competencies, as language, regional expertise, diplomacy, and social etiquette.[1] For example, in his 2006 testimony to the Senate Armed Services Personnel Subcommittee, then–Air Force Deputy Chief of Staff for Personnel Lt Gen Roger A. Brady stated: "Today's dynamic security environment and expeditionary nature of air and space operations require a cadre of Air Force professionals with a deeper international insight, foreign language proficiency, and cultural understanding" (Brady, 2006, p. 5). And Lt Gen Stephen R. Lorenz, the Air Force's Air University Commander, said: "[A]s the United States begins to understand the nature of the long war, the need for training in language and regional cultures has become even more apparent" (Lorenz, 2007). Similar sentiments on the importance of these types of cross-cultural skills have been expressed by many other officials, including Secretary of the Air Force Michael W. Wynne (in a Letter to Airmen [Wynne, 2006]), both President George W. Bush and Secretary of State Condoleezza Rice (in remarks made at the U.S. University Presidents' Summit on International Education [Bush, 2006; Rice, 2006]), and Deputy Secretary of Defense Paul Wolfowitz (in an interview with the *New York Times* [Wolfowitz, 2003]).

The Department of Defense (DoD) has codified the importance of cross-cultural training in several official programs and documents, including the National Security Language Initiative (U.S. Department of State, 2006), the Quadrennial Defense Review Report (DoD, 2006), and the Department of Defense Language Transformation Roadmap (DoD, 2005). Moreover, the Roadmap mandates that every service should develop and track the language and regional expertise of its personnel, and clearly states that language, culture, and regional expertise are not only important "defense core competencies," but also "critical weapons systems."

Given the presence of U.S. forces in numerous countries, particularly in current operations, the Air Force and other services have begun taking steps to integrate cross-cultural training into existing curricula. This call for cross-cultural skills has led Air Force leaders to acknowledge the need for much improvement in the Air Force's training of cross-cultural performance.

[1] Although the Air Force clearly acknowledges that cross-cultural performance is important, it has not yet provided a clear definition of what cross-cultural performance is.

Defining the Problem

In response to the DoD guidance for all services to improve the cross-cultural performance of their forces, the Air Force Directorate of Airman Development (AF/A1D) asked RAND to help conceptualize the content of training programs aimed at improving cross-cultural performance within the Air Force. We understood that the Office of the Secretary of Defense (OSD) and the Air Force were primarily concerned with the application of cross-cultural performance skills, or competencies, in the field. Accordingly, we focused on the use of *training* to improve cross-cultural performance and use the term *training* in this document to refer to any efforts intended to improve job performance. We recognize that efforts labeled as *education* (e.g., professional military education, or PME) also play a role in improving job performance.[2] Thus, even though we focused on training, the results of our analyses also apply to aspects of education programs.

To accomplish the project's goal, RAND researchers began by setting up informal focus groups, interviews, and meetings with various Air Force personnel in order to gain an initial understanding of the types of cross-cultural training needed (the demand for training) and the types of training currently available (the current supply of training).

Through these informal interviews and focus groups, the researchers discovered that despite leadership's clear agreement that cross-cultural performance is important, there was little agreement on the type of training needed to improve cross-cultural performance. For example, some of the Air Force trainers and instructors who provide cross-cultural training think that language is the key to improving cross-cultural performance. Others think that regional education, including geography, history, political information, and economic information, will address the need for cross-cultural skills. Still others think that teaching regional norms, social etiquette, taboos, etc., will fill the cross-cultural training gap. In many cases, advocates for one type of training are adamantly opposed to the need for another. In addition, advocates for many existing training programs argue that they are already providing cross-cultural training, even though the content of the training differs noticeably from program to program. From our conversations with various Air Force personnel and Air Force trainers and educators, it became clear that people were requesting a wide variety of training and that the types of training already being provided differed widely from course to course. This led us to conclude that the type of cross-cultural training being requested might not be consistent with the type of training being provided.

Without consensus on what constitutes cross-cultural training, there is no way to know whether cross-cultural training needs are being met by every or even any existing training program. More specifically, lack of consensus on the relevant components of a cross-cultural training curriculum could result in a disconnect between the content of the training and the needs expressed by airmen returning from deployment. For example, if an airman requesting cross-cultural training is looking to be trained in social etiquette and receives training that consists of information on history, economics, and political systems, that airman's training needs are not being met.

Without first establishing what type of training airmen need, one cannot be certain whether that need is being met by existing programs. For this reason, the critical first step is to

---

[2] We use *training* and *training programs* as an inclusive term to cover training, education, experience, and development efforts (Chapter Two elaborates on this usage).

define the domain of cross-cultural performance so that it can be used to guide the development of training that meets Air Force performance needs and ensures effective use of training resources.

Our discussions with various Air Force personnel also highlighted the disjointed and unsystematic nature of current Air Force–wide cross-cultural training efforts. Many commendable and well-intentioned training programs exist in the Air Force, but their availability is limited, particularly for airmen who may need them the most. For example, some Officer Training School (OTS) courses cover regional education (i.e., geography, history, political information, and economic information on major regions of the world) but are not accessible to all airmen who may need this type of training. A computer-based language training program that offers flexible training times and broad accessibility via the Internet is another example. It is available to only some airmen, and many of those are unaware of its availability or may not have time for the training.

Based on what we learned in our informal interviews and focus groups with various Air Force personnel, we identified several other gaps in the Air Force's current cross-cultural training efforts. One of these is the absence of an established method for evaluating the success of existing cross-cultural training; another is a limited capability for tracking how much and what type of cross-cultural training individual airmen receive. Except in the case of language, the Air Force's capability for measuring or accounting for the existing cross-cultural skills of its force is limited.

A good starting point for closing these gaps is a systematic inventory of the Air Force's cross-cultural training objectives. A comprehensive examination of all possible definitions of cross-cultural performance is needed to

- facilitate communication between those requesting cross-cultural training (the demand side) and those providing it (the supply side)
- assess airmen's specific training needs
- provide targeted training and education to meet those needs
- provide basic criteria to evaluate each type of training's success at meeting all or any relevant training needs.

Our Approach

To better understand the Air Force's cross-cultural training needs, we conducted a basic needs assessment for cross-cultural training to establish what and how much of particular types of behavior are required for improved cross-cultural performance—a process conspicuously absent in much of the research literature on cross-cultural training (Campbell and Kuncel, 2001). Our intention was not to define *culture*, but, rather, to define what airmen should be able to do in a foreign culture. This information can serve as the foundation for the Air Force's establishment of training objectives, which is the essential first step in creating a systematic program of cross-cultural training.

We approached the problem through three main questions:

1. What is cross-cultural performance, or behavior?

2. Which cross-cultural behaviors do airmen identify as important to their deployed jobs?
3. Do all airmen, regardless of job requirements, need the same type and/or amount of cross-cultural training?

We began by first reviewing the existing literature on cross-cultural training and performance and holding discussions with various Air Force personnel to determine what airmen need to be able to do to be considered "cross-culturally competent." We then used what we learned in the review and discussions to develop 14 categories of cross-cultural behavior that are potentially relevant for on-the-job cross-cultural performance. Chapter Two provides a full discussion of what went into addressing question 1.

Next, we tested the relevance of the 14 behavior categories by surveying approximately 21,000 previously deployed airmen and asking them to rate the importance of these categories in their deployed job. Consistent with standard methods of job analysis (Goldstein, 1991; Williams and Crafts, 1997), these importance ratings were used to determine whether the 14 behaviors can be considered an important component of job performance for a particular job. See Chapter Three for a full discussion of how we addressed questions 2 and 3.

We also used the results of our survey to arrive at suggestions for training components and to describe key steps remaining in the development of a comprehensive program for bolstering the cross-cultural competence of airmen—such as skills assessment, evaluation of existing training programs, and development of training for specific deployment locations. Chapter Four provides a full discussion.

CHAPTER TWO

Defining Cross-Cultural Job Performance

As discussed in the previous chapter, our goal was to determine which behaviors are necessary for successful cross-cultural performance in the Air Force. We began by searching the existing literature for frameworks to use in defining cross-cultural performance. What we found is that there is a substantial body of work on theoretical frameworks for cultural differences, specific types of cross-cultural training for certain behaviors, and expatriate and sojourner experiences. However, there is little work defining the full domain of cross-cultural performance.

Examining Frameworks for Differentiating Cultures

The theoretical work on frameworks for comparing and describing cultures is extensive. These frameworks are used to define the dimensions of culture, or the constellations of shared values, norms, and beliefs that characterize culture. Though the exact definitions and dimensions of culture are still subject to debate (Triandis, 1996), examples of dimensions commonly used to describe cultures include individualism and collectivism (the extent to which persons define themselves in terms of their personal characteristics and goals versus the characteristics and goals of collectives to which they belong), verticality and horizontality (the extent to which cultures value hierarchical versus egalitarian relationships), and masculinity and femininity (the extent to which cultures value assertiveness) (Triandis, 1996; Triandis and Bhawuk, 1997; Hofstede and Hofstede, 2005).

While these culture typology approaches have given rise to their own substantial bodies of literature, they focus on broad-brush categorization systems by which cultures can be differentiated. Thus, this literature primarily seeks to define *culture* itself, not to determine behaviors essential for successful job performance in a culture. Descriptions of the dimensions of culture are simply too broad to help clarify the Air Force's training objectives or to help specify the behaviors required when working and living in a foreign culture.[1]

A number of researchers have applied these theoretical culture-defining approaches to the training context (for example, Bhawuk, 1998; Fiedler, Mitchell, and Triandis, 1971; Worchel and Mitchell, 1972). However, these efforts typically aim not to improve job performance per se, but to bring about a better understanding of differences between cultures. Despite the presumption that an awareness of these differences leads indirectly to improved performance, the

[1] Definitions of culture and typologies used to describe and compare cultures can, of course, be very useful in helping students in cross-cultural courses to understand and apply course material. For further discussion of this issue, see Chapter Two's section on additional issues for consideration.

effectiveness of current training interventions based on these approaches has not been thoroughly investigated.

The Lack of Cross-Cultural Performance Training Evaluations

Not only is there no universally agreed upon definition of culture, there is also, as mentioned earlier, no agreement on what constitutes cross-cultural performance. The literature examining this performance often uses broad, non-performance-based criteria (Morris and Robie, 2001; Ones and Viswesvaran, 1997), such as withdrawal or early termination of assignments (e.g., Abbe, Gulick, and Herman, 2007; Bhaskar-Shrinivas, Harrison, Shaffer, and Luk, 2005; Birdseye and Hill, 1995). Of the studies that focus on performance abroad, many tend to draw on existing, well-validated general models of domestic task and contextual performance (e.g., Caligiuri, 2000; Shaffer, Harrison, Gregersen, Black, and Ferzandi, 2006) and pay little attention to what is unique to cross-cultural performance.

These broad theoretical approaches to job performance are too general to be of much use in creating training programs (Campbell and Kuncel, 2001) because they provide no performance information at a level of detail sufficient for determining specific training objectives—in this case, the specific behavioral components of cross-cultural performance.

The operationalization and subsequent measurement of behaviorally based (and hence potentially trainable) cross-cultural competencies is at a relatively rudimentary stage (Dinges and Baldwin, 1996). Even a broad attempt by Arthur and Bennett (1995) to determine the importance of 54 possible characteristics for expatriate performance examined relatively few that addressed specific behavior. Expatriates were asked to rate the importance to success of a list of factors ranging in behavioral specificity from relatively precise (such as courtesy and tact, display of respect) to even less precise (such as high motivation, youthfulness, positive self-image).

Researchers in cross-cultural performance recognize that a needs assessment is a key prerequisite for the development of an effective cross-cultural training program (Gudykunst, Guzley, and Hammer, 1996), but there is little evidence of any comprehensive investigation of the full set of potentially necessary cross-cultural competencies. Moreover, much of the research has been conducted in a civilian rather than a military context. Given the differences in the goals and duties of civilian jobs versus military missions, this constitutes yet another barrier to these studies' usefulness to Air Force cross-cultural performance training.

Lack of empirical validation is also a problem in the literature on cross-cultural performance training (Black and Mendenhall, 1990; Church, 1982; Mendenhall, Stahl, Ehnert, Oddou, Osland, and Kuhlmann, 2004; Morris and Robie, 2001). As noted previously, much of this literature assumes that training in certain competencies is essential for successful cross-cultural performance, thus failing to fully recognize that this is an empirical question that has gone largely untested. For example, some of the studies look at specialized training programs, called "cultural assimilators" (Bhawuk, 1998; Fiedler, Mitchell, and Triandis, 1971; Worchel and Mitchell, 1972), that are intended to train individuals to correctly attribute motivations for behavior in different cultures (Cushner and Landis, 1996). These studies assume that such training will facilitate cross-cultural interaction but do little to confirm that making correct attributions is, by itself, sufficient to enhance performance.

Another study describes the development of a cross-cultural training program focused on cross-cultural communication (Cushner and Brislin, 1996). There is theoretical and logical support for considering communication a component in the constellation of necessary skills, just as there is for cultural attributions. However, in this case too, there has been no empirical validation of communication training's necessity, let alone sufficiency, for improving job performance. Efforts on other specific components of cross-cultural performance, such as negotiation (e.g., Ting-Toomey, 2004), also suffer from this problem.

Thus, just as there are differences of opinion among Air Force subject-matter experts about cross-cultural performance training, there is little agreement in the literature on the comprehensive scope of cross-cultural behavior training. However, the existing literature does provide a good starting point for exploring the full constellation of behavioral competencies needed for successful cross-cultural performance. That literature, in combination with informal focus groups and discussions with various Air Force personnel, helped us identify our 14 categories of cross-cultural behavior.

The 14 Cross-Cultural Behavior Categories

Our review of the existing literature and discussions with various Air Force subject-matter experts made it clear that the terms *cross-cultural job performance* and *cross-cultural job skills* brought to people's minds a wide range of behaviors and that those behaviors differed considerably from individual to individual. This meant that before a training program to improve cross-cultural job performance could be designed, we would have to define the domain of behaviors to be trained.

As there was no established classification system covering all potentially relevant aspects of cross-cultural performance, we developed one. This taxonomy consists of 14 cross-cultural behavior categories. Several of these categories overlap conceptually, but we chose to use all 14 because each one had been deemed important by at least some Air Force personnel and/or research literature. One benefit of this taxonomy is that it provides a set of behaviorally specific terms, use of which can improve communication between those who supply and those who request cross-cultural performance training.

We grouped the 14 categories according to two global types of behaviors: enabling and goal oriented. *Enabling behaviors* are those that help facilitate a variety of day-to-day activities and are likely to be needed in a variety of jobs. *Goal-oriented behaviors* are those associated with specific mission-related activities and are likely to be needed only by individuals in certain Air Force Specialty Codes (AFSCs) or for certain job tasks. Successful performance of enabling behaviors will likely improve one's performance of goal-oriented behaviors but is not a prerequisite for successful performance of goal-oriented behaviors. Individuals could, in theory, have low skills in an enabling behavior (e.g., foreign language) and yet have high skills in a goal-oriented behavior (e.g., negotiating with others).

Enabling Behaviors

Foreign language skills. The foreign language skills category includes the abilities to speak, write, read, and understand a non-English language. For Americans visiting a non-English-speaking country, even a small amount of language capability offers an advantage in interacting with the locals and the culture. Understanding and speaking the language can demonstrate respect for, interest in, and willingness to make the effort to learn about the locals, and can facilitate more-personal interactions and eliminate misunderstandings that can occur when an interpreter is involved. The ability to read street signs, local newspapers, etc., can facilitate living and working in the local community. Understanding the language can also help improve the visitor's interpretation of interactions with and between locals; speaking the language permits greater immersion in the culture, which leads to increased opportunities for deeper learning about the culture. Because of how much language skills offer, they should be viewed as a key aspect of cross-cultural performance.

Research on non-native speakers suggests that language skills can have both a positive and a negative effect on locals' perceptions. Several studies show that higher levels of fluency in a foreign language result in more-positive perceptions by native speakers (White and Li, 1991; Wible and Hui, 1985; Molinsky, 2005) and higher job performance (Mol, Born, Willemsen, and Van Der Molen, 2005). However, other research shows that as a person's fluency in a foreign language increases, so do the expectations that the person's behavior will be consistent with interpersonal cultural norms (Molinsky, 2005). In other words, culturally inappropriate behavior is more acceptable from a foreign person whose language fluency is low. This suggests that even though language fluency can be an important skill for increasing locals' positive perceptions, it may be advisable to accompany this skill with other culturally relevant skills in order to avoid producing a highly fluent individual who is viewed as culturally inappropriate.

Verbal and nonverbal communication. Verbal and nonverbal communication is the sending and understanding of spoken and unspoken information about mood, intent, status, demeanor, and message (Cushner and Brislin, 1996; Cushner and Landis, 1996; Gudykunst, Guzley, and Hammer, 1996). This is not the same as foreign language skills: Native speakers of a language can be poor communicators, and people with no language skills can communicate effectively with locals in a foreign country.

We included this behavior category to distinguish specific language skills from general communication skills in the domain of cross-cultural behavior, since we had observed that the two were often conflated in Air Force personnel's discussions of cross-cultural training needs. The literature on cross-cultural training notes the importance of both verbal and nonverbal communication (e.g., Cushner and Brislin, 1996; Cushner and Landis, 1996).

Applying appropriate social etiquette. The application of appropriate social etiquette entails knowing and being able to use the etiquette—customs, conventions, norms, manners, traditions, gender-specific rules, etc.—called for in a given situation in a given location. Each country and region has its own social etiquette for different contexts. For example, in the United States, appropriate social etiquette at a black-tie dinner differs from that at a sports game, in a corporate boardroom, or on the streets of Manhattan. The importance of skills in this category is supported by the literature (Arthur and Bennett, 1995).

Managing stress in an unfamiliar cultural setting. Stress management in an unfamiliar cultural setting entails being able to recognize and deal with stress resulting from a lack of language skills, an inability to understand cultural norms, feeling like a "fish out of water" because of cultural and ethnic differences, feelings of being perceived by locals as a stereotypical American, and feeling that locals may not welcome the presence of U.S. troops.

Expatriates may feel stress and anxiety during interactions with people from a different culture or in an unfamiliar cultural setting (Cushner and Brislin, 1996; Black and Mendenhall, 1990; Black, Mendenhall, and Oddou, 1991). It is important for expatriates to learn to manage these reactions, because stress and anxiety that are not managed and/or reduced can result in increased illness, absenteeism, lower job performance, lower job satisfaction, and higher turnover (Bhaskar-Shrinivas, Harrison, Shaffer, and Luk, 2005; Cooper, Dewe, and O'Driscoll, 2001; Hom and Griffeth, 1995). Psychological adjustment has been studied extensively in the expatriate literature as an outcome measure of expatriate success (e.g., Ward and Kennedy, 1993; Janssens, 1995; Ward, Okura, Kennedy, and Kojima, 1998; Shaffer, Harrison, Gregersen, Black, and Ferzandi, 2006; Cross, 1995) and has been shown to affect expatriate job performance (Bhaskar-Shrinivas, Harrison, Shaffer, and Luk, 2005). Emotional stability, a personality trait related to how one deals with stress, has also been shown to relate to expatriate job performance (Mol, Born, Willemsen, and Van Der Molen, 2005).

Changing behavior to fit cultural context. Changing behavior to fit cultural context involves adapting one's behavior to match and/or complement the behavior of locals, avoiding common American practices that may be offensive in certain settings, and behaving in ways that might be awkward for an American but are necessary to fit in with the culture. This category is similar to that of applying appropriate social etiquette (described above) except that, here, there is an emphasis on modifying one's behavior to match the situation.

Gathering and interpreting observed information. The gathering and interpreting of observed information consists of watching locals interact both with each other and with U.S. military personnel to learn what produces positive and negative reactions in both cases. This category also entails observing one's surroundings with the objective of learning to identify which locals appear to be acting inappropriately or suspiciously and which do not.

Bandura's (1977) theory of social learning suggests that opportunities for observation and modeling may increase an individual's learning about proper modes for interaction in a foreign culture (Mak, Westwood, Ishiyama, and Barker, 1999). As a person's opportunities for observation in a culture grow, so too should the person's skill at interacting within that culture. However, foreigners in an unfamiliar cultural environment often avoid opportunities for observational learning because they find interacting with the locals anxiety inducing (Mak, Westwood, Ishiyama, and Barker, 1999). It is thus suggested that training in cross-cultural behavior should include instruction on how to observe and interpret interactions between locals.

Applying regional knowledge. The application of regional knowledge entails knowing and being able to use factual information about a region or country's economic, political, and religious history, as well as its current economic, religious, and political issues. It also entails using factual information about relevant foreign relations (e.g., whether a neighboring country is an ally or foe).

Knowledge about a region or country provides a basic foundation for any involvement in that culture, and understanding how to apply that knowledge to strategic decisionmaking is likely a critical skill for many military personnel serving in a foreign culture. Also, basic knowledge about a region can demonstrate one's interest in and willingness to put effort into learning about the culture. Bhawuk (1998) and others have found that it is important for all personnel involved in a foreign culture to learn how to apply at least some regional knowledge in their day-to-day jobs.

Self-initiated learning. Self-initiated learning is defined as taking the initiative to learn more about the country, culture, or language than was provided in one's training. It requires a willingness to go above and beyond that training and a desire to constantly improve one's cross-cultural performance. No training program will ever be comprehensive enough to preclude additional learning, so this is a way for individuals to continue to augment their training and improve their performance while deployed. It can take the form of volunteering for or requesting additional training from the Air Force; spending off-duty time with locals; talking with someone from the country, such as an interpreter or guide; and learning on one's own through the Internet, books, or computer-based training software.

Self-initiated, proactive learning is not typically included in the existing cross-cultural training literature (although Arthur and Bennett, 1995, suggest that initiative is important), but we think that this category of behavior is an integral component of cross-cultural performance. Theoretically, high performance in self-initiated learning would demonstrate to the locals that one is interested in the culture and willing to make the effort to learn about it, which may result in the locals becoming more receptive and welcoming to proactive learners. At the same time, high performance in this behavior category may help improve performance in all other cross-cultural behavior categories.

Respecting cultural differences. Respecting cultural differences consists of respecting locals' values, opinions, and mindset, and recognizing that stereotypes—which every culture has—should be avoided and not used as the basis for opinions. This category also includes recognizing that American culture differs from other cultures but is not necessarily superior, and demonstrating respect for cultural differences between Americans and locals.

Cultural sensitivity has been acknowledged as an important aspect of cross-cultural performance (Hammer, Bennett, and Wiseman, 2003); it has also been shown to relate to job performance (Mol, Born, Willemsen, and Van Der Molen, 2005). The relationship between this behavior category and job performance is not surprising given the ease with which people form and accept stereotypes about members of other groups and the detrimental effects of stereotypes on interpersonal interactions. If airmen are expected to interact with individuals from other cultures, it is advisable that they learn to respect cultural differences, which includes avoiding stereotypes.

Goal-Oriented Behaviors

Establishing authority. Establishing authority covers actions needed when exercising judicial and/or law enforcement powers, controlling or restricting the behavior of locals, ordering compliance from locals, or demonstrating that one is in charge. This category of behavior would likely be needed by individuals serving in a law enforcement or security forces capacity.

Research has demonstrated that cultures vary in their reactions to different techniques for exerting authority (Tyler, Lind, and Huo, 2000). Establishing authority thus may be an important behavior category for inclusion in cross-cultural performance training for certain jobs.

Influencing others. Influencing others involves changing locals' opinions or behavior, convincing them to follow one's leadership willingly (i.e., without force or coercion), providing them with guidance or leadership, persuading them to go along with or accept an idea, and influencing or persuading them to behave or act in a certain way. This category differs from that of establishing authority (directly above) because the emphasis is on convincing locals to voluntarily follow one's leadership and guidance rather than on making them comply with one's orders.

High performance in this behavior category leads to greater backing and support by locals and, ultimately, to greater success at motivating locals to assist in achieving a performance goal.

This category is similar to the leadership/management behaviors described as important in some cross-cultural training literature (e.g., Harrison, 1992). Effective leadership/management behaviors (defined as behaviors successful at exerting influence) have been shown to differ from culture to culture (Gelfand, Erez, and Aycan, 2007). For example, managerial behavior that predicts managers' performance in the United States has been shown to be unrelated to either American or Chinese managers' performance in Hong Kong (Black and Porter, 1991).

Moreover, motivators and rewards differ across cultures (Gelfand, Erez, and Aycan, 2007), which means that behaviors for influencing others should take into account culturally relevant motivators and rewards. Individuals working in other cultures must understand the methods that are and are not persuasive in a given culture.

Negotiating with others. Negotiating with others entails, among other skills, bargaining successfully with locals for supplies or resources, reaching compromise solutions with locals that please both sides, and forming mutually beneficial partnerships with locals.

Effective negotiation tactics differ from culture to culture (Brett and Okumura, 1998; Sheer and Chen, 2003; Gelfand, Erez, and Aycan, 2007). Failure to use the appropriate tactics may result in less negotiating power, loss of respect, and in some cases failed negotiations. It is therefore imperative that airmen working in other cultures understand the methods that are and are not appropriate and effective in a given culture.

Establishing credibility, trust, and respect. Establishing credibility, trust, and respect means improving the locals' perception of and respect for Americans and the U.S. military and showing them that Americans will not betray their trust. Gaining credibility and the trust and respect of the locals can be important not only for accomplishing the immediate mission, but also for improving the locals' perceptions of Americans and the U.S. military in general and thereby helping to ensure the success of future missions.

Methods for establishing trust can vary across cultures (Sullivan, Peterson, Kameda, and Shimada, 1981), and people often automatically view members of groups other than their own as less trustworthy, credible, and deserving of respect. In addition, failure to establish credibility, trust, and respect can impede normal business interactions (Doney, Cannon, and Mullen, 1998).

To have successful interactions with locals, airmen need to demonstrate that they are worthy of trust and respect by striving to be a credible source of information before proceeding with business. This is particularly important in a military context, since many locals may feel threatened by the presence of the U.S. military or may distrust U.S. intentions.

Resolving conflict. Resolving conflict involves preventing, managing, defusing, and otherwise resolving conflicts between locals, as well as between Americans and locals.

Conflict resolution skills are important for airmen who interact with locals or manage personnel who interact with locals. Conflict reduction is probably most needed by airmen who manage local people, assist a community in a policing capacity, or serve in a diplomatic role. It is important that the differences in conflict resolution approaches used by different cultures be understood (Tyler, Lind, and Huo, 2000; Sullivan, Peterson, Kameda, and Shimada, 1981; Ohbuchi and Takahashi, 1994; Morris, Williams, Leung, Larrick, Mendoza, Bhatnagar, Li, Kondo, Luo, and Hu, 1998; Gelfand, Erez, and Aycan, 2007).

Additional Issues to Consider

In developing a comprehensive program for cross-cultural performance training, several additional concepts about cross-cultural training are worth noting; therefore, we expand on the meaning and relevance of these terms here.

Defining Culture During Training

Definitions and taxonomies of culture are likely to be useful as teaching tools in certain cross-cultural performance courses. Defining culture as a shared set of norms, beliefs, expectations, values, etc., and demonstrating how to apply that definition in the context of the 14 categories of cross-cultural behavior could improve students' interpretation of locals' actions in a specific context. For example, in the context of negotiation training, understanding the norms and expectations for negotiation and what people value (interpersonally and monetarily) would likely facilitate effective negotiation.

No single definition of culture can fully describe all relevant aspects of a culture; moreover, it would be difficult to locate a single specific definition that meets with universal approval. Instead, more-specific definitions are likely useful for some purposes but not for others. For this reason, we suggest that the Air Force consistently, and broadly, define culture as a shared set of norms, beliefs, expectations, values, etc. A definition of this breadth captures the commonalities in many existing definitions of culture (Triandis, 1996). A variety of more-specific, course-relevant operationalizations of culture (such as Hofstede's taxonomy) may then be introduced as needed to aid training. Some training applications of specific operationalizations of culture are available, but the issue of which operationalizations are most applicable to which training topics remains relatively unexplored, and the usefulness of different definitions for particular training domains should be further investigated. Additionally, because trainees may encounter more than one specific operationalization of culture across the courses they take, instructors should regularly clarify the fact that there is no single definition of culture that can describe all relevant aspects of a culture and that Air Force training courses thus may use different, course-specific definitions.

Training, Education, Development, and Experience

We are using the term *cross-cultural training* to mean the effort to prepare personnel for the cross-cultural performance requirements of current and future missions. *Training* is often used interchangeably with such terms as *education*, *development*, and *experience*. For example, when students discuss their graduate course work, they often make no distinction between graduate *training*, graduate *experience*, and graduate *education* (e.g., "During my graduate training, I learned about. . . ."). The military, however, has established distinctions between *training*, *education*, *experience*, and *development*.

According to *Officer Professional Military Education Policy (OPMEP)*, Chairman of the Joint Chiefs of Staff Instruction (CJCSI) 1800.01C (Joint Chiefs of Staff, 2005), p. A-1–A-2:

> a. Professional development is the product of a learning continuum that comprises training, experience, education, and self-improvement. PME provides the education needed to complement training, experience, and self-improvement to produce the most professionally competent individual possible.

b. In its broadest conception, education conveys general bodies of knowledge and develops habits of mind applicable to a broad spectrum of endeavors. At its highest levels and in its purest form, education fosters breadth of view, diverse perspectives and critical analysis, abstract reasoning, comfort with ambiguity and uncertainty, and innovative thinking, particularly with respect to complex, non-linear problems. This contrasts with training, which focuses on the instruction of personnel to enhance their capacity to perform specific functions and tasks.

c. Training and education are not mutually exclusive. Virtually all military schools and professional development programs include elements of both education and training in their academic programs. Achieving success across the joint learning continuum relies on close coordination of training and education to develop synergies as personnel develop individually over time, acquiring and performing progressively higher skills and responsibilities as their careers advance.

These three statements make it evident that distinctions among training, education, experience, and development can be and are made; they also make it evident that there is a great deal of overlap among these elements, all of which are relevant for preparing personnel to perform their missions. For these reasons, and for the sake of simplicity, we thus make no distinctions in this report. We use the term *cross-cultural training* to encompass all of what elsewhere may be referred to as cross-cultural education, cross-cultural development, and cross-cultural experience, as well as cross-cultural training.

Training Occurring Over an Airman's Career and Just Prior to Deployment

We distinguish between *over-career training*, which occurs across an airman's career (and includes training geared toward providing continuous or life-long learning opportunities), and *predeployment training*, which an airman receives just prior to being deployed. Cross-cultural performance involves behavior that cannot be fully learned solely through predeployment training. Ideally, this learning should start during basic training, the goal being to provide airmen with a basic foundation of skills in the 14 categories and to emphasize the strong value that the Air Force places on cross-cultural performance. The basic foundation should then be developed further in over-career training courses (e.g., PME courses) covering the 14 categories in greater depth.

The benefits of cross-cultural training can easily be lost if such training is but one of many elements of predeployment training (others include safety training, combat training, and base procedures training). Moreover, predeployment training works best not as a primary means of providing training, but as a refresher course and as a vehicle for covering critical and previously unforeseen issues.

Culture-General and Culture-Specific Training

Cultural-general training refers to training that provides understanding of the myriad ways in which cultures can and do differ. Culture-specific training refers to training for a specific country, region, or social group. Culture-general training is not synonymous with over-career training; and culture-specific training is not synonymous with predeployment training. Rather, culture-specific training can occur in both over-career and predeployment training. Similarly, cultural-general training can occur in both over-career and predeployment training.

Both culture-general and culture-specific training should be included in all cross-cultural training courses, as the understanding of one type of material likely aids the understanding of the other type. Understanding what is taught in culture-general training (which should include the definition of culture itself and the defining characteristics of cultures) can serve as a platform for understanding culture-specific course material. It can ensure that Air Force personnel who encounter a specific culture have a solid understanding of the types of differences to expect, how to interpret those differences, and possible reasons for them.

Though some, but not all, over-career training should be culture general to establish skills that apply across cultures, predeployment training should clearly include culture-specific training. Such training will likely be relatively brief, requiring curricula that are sharply focused on the specific and immediate needs of the airmen being deployed.

Training for Current Missions and Future Missions

As noted previously, cross-cultural training should prepare airmen for not only current missions, but also future missions. Offering all airmen over-career training that allows them to specialize in one of a variety of cultures would ensure that specialists are available for any region should a need arise.

Another way to prepare the force for future missions is to make sure that past learning and areas of specialty are maintained as part of over-career training. Previously learned knowledge and skills can degrade over time, particularly if there is no opportunity to practice or apply them. This is especially relevant for cross-cultural performance, because an airman's opportunities to engage with a specific culture do not naturally present themselves unless he or she is living in that culture. Refresher courses and/or periodic re-certification of knowledge and skills are therefore an important part of ensuring future mission readiness.

Predeployment training is clearly one way to fill the need for immediate, additional training for a current mission. However, many of the 14 categories of behavior cannot be fully developed or even introduced in a few days of predeployment training. For that reason, continuous training is the advisable way to ensure that current mission needs are met. Even for airmen who have a solid foundation built on prior cross-cultural performance training, refresher courses are vital to ensuring that immediate mission needs are met.

CHAPTER THREE

Cross-Cultural Performance Survey

After creating our list of 14 behavior categories of cross-cultural performance needed in a foreign environment, we designed a survey to confirm the importance of these categories for deployed jobs in the Air Force and to examine airmen's perceptions of the helpfulness of existing training in the 14 categories. This chapter describes the goals of the survey, the method used, and the results of our analyses.

Goals

Our survey had four primary research goals:

1. Verify the importance of the 14 identified categories of cross-cultural behaviors.
2. Identify any remaining categories of cross-cultural behaviors not covered by the existing 14.
3. Determine whether training needs differ by AFSC, grade (enlisted and officer), and deployment location.
4. Determine how helpful current Air Force training in these behavior categories is perceived to be.

Method

Participants

Because our goal was to identify the types of behaviors important for deployed jobs, we decided to survey recently deployed airmen (i.e., airmen who had deployed or returned from deployment in the previous 18 months) about their job during deployment. In this way, we would obtain responses that were not only from knowledgeable sources, but also based on first-hand experience. This method, surveying recent job incumbents, is consistent with typical needs assessment techniques (e.g., Goldstein, 1991).

To examine differences in training needs across AFSCs, grades (enlisted personnel and officers), and deployment locations, we developed a complex sampling plan that used disproportionate random stratified sampling to ensure adequate sample sizes within each of the subgroups of interest. These stratified subgroups were defined by the following demographics:

- two-digit AFSC
- grade: low (E-1 through E-4 for enlisted; O-1 through O-3 for officers) versus high (E-5 through E-9 for enlisted; O-4 through O-9 for officers)
- deployment location: Iraq/Afghanistan versus other.

Information on AFSCs, grade, and deployment location for all deployed personnel were supplied by the Air Force Personnel Center and subsequently used to select our sample.

To estimate the required sample size per subgroup, we estimated the target number of respondents and then increased it to accommodate the expected survey response rates. Approximately 60 respondents per subgroup are necessary for 80 percent power to detect a difference of .5 standard deviations between two groups (we reasoned that differences smaller than .5 would not be of practical significance in designing or evaluating training programs). Because past experience with Air Force surveys led us to expect a 30 to 50 percent response rate, we increased the number of those we would invite to participate per subgroup to 150. For all subgroups whose population was larger than 150, we invited a random sample of 150 people. For subgroups whose population was smaller than 150, we invited the entire subgroup. Appendix A displays the population size and number of invitees per subgroup.

Approximately 23,000 individuals were selected for participation in the survey. Our initial contact with these individuals was an email, sent by the Air Force Manpower Agency, inviting them to take part in the survey. Email addresses of 1,174 individuals (5 percent of the 23,000) proved to be invalid, which left us with a final total of 21,846 email invitations sent. Follow-up emails were then sent to individuals who had not responded within a specific time.

Of the 21,846 who received email invitations, 8,498 (39 percent) logged on to the survey, and 6,653 (30 percent) completed at least 50 percent of all survey items. The total number of respondents for each item ranged from 6,272 to 7,757.

Respondents reported a wide variety of locations for their most recent deployment, the three most common being Iraq (35.7 percent), Qatar (18.1 percent), and Afghanistan (12.3 percent). Other deployment locations reported by at least 1 percent of the sample included Kuwait (7.1 percent), the United Arab Emirates (UAE) (4 percent), Kyrgyzstan (3.1 percent), Germany (2.3 percent), Korea (1.5 percent), Djibouti (1.1 percent), Guam (1.1 percent), and Diego Garcia (1 percent). Some individuals (4.6 percent) indicated that their deployment location was classified or otherwise did not provide an answer. Of the entire sample, 62.5 percent reported that they were enlisted, 36.6 percent reported that they were officers, and the remainder (0.9 percent) did not disclose their grade.[1]

Survey Measures

The survey used the following measures (Appendix B contains a complete copy of the online survey instrument):

Demographics. The first part of the survey asked respondents such questions as the location of their most recent deployment and the AFSC and grade of the job in which they were deployed.

Importance of the 14 behavior categories. We developed 70 items to evaluate airmen's perceptions of the importance of our 14 behavior categories (three to six items per category).

---

1 All percentages reported here are unweighted. Some demographics are overrepresented because of the disproportionate stratified sampling.

These items cover a range of on-the-job behaviors specific to each category, thus creating a scale (a group of items used to measure a single concept) for each category. For example, for the category of changing behavior to fit cultural context, we developed four items: (1) "Adapt my behavior to match and/or complement the behavior of local people," (2) "Change or deviate from SOPs [standard operating procedures] and/or ROE [rules of engagement] to accommodate cultural issues," (3) "Interact with locals in ways that might be awkward for an American but necessary to fit in with the culture," and (4) "Avoid common American practices that may be offensive in certain cultural settings, such as speaking loudly, pointing, or using the left hand to pass an object." (For a list of the items for each scale category, see Appendix C.)

In this section of the survey, participants were asked to indicate their level of agreement (on a rating scale from strongly disagree = 1 to strongly agree = 5) with 70 statements that each began with the prompt "In ______, [participant's most recent deployment location was inserted here] I found myself in situations (on and off duty) where it was or would have been important for me to:" and ended with each of the 70 items.[2] The survey presented the 70 items in random order.

We computed an importance score for each of the 14 categories. It is a scale score, computed as the average of the importance ratings on the three to six individual items for each category. Only the participants who completed all items in a category's scale were included in the score calculation for that scale. Participants who selected "Don't know" for one or more items were not included.[3]

We then identified three (out of the 14) scale scores (those for applying appropriate social etiquette, self-initiated learning, and changing behavior to fit cultural context) for which coefficient alpha increased when a single item was removed;[4] that led us to remove the item in all three cases and recompute the scores for those scales. The order of importance for the categories changed when these items were excluded, so we decided to report results for both the original scale and the "shortened" scale.[5] (Appendix D provides the weighted summary statistics—weighted correlations, descriptive statistics, and coefficient alphas—for the importance ratings of the 14 behaviors on both scales.)

Perceptions of training. The section on training asked participants to indicate their agreement with statements that began with "The training I received from the Air Force just prior to deployment in ______ helped me to . . ." and "When I was deployed in ______, the training I received from the Air Force over the course of my career helped me to . . . ," and ended with each of the 14 cross-cultural behaviors. Participants responded to each statement using a scale

---

2 The prompt includes the phrase "on and off duty" because Air Force performance evaluations generally regard both on-duty and off-duty conduct of Air Force personnel as important. Moreover, the inclusion of off-duty conduct is particularly important because all of the behavior of Air Force personnel in a foreign country can influence how the Air Force and the United States are perceived. Our definition of Air Force cultural performance therefore includes both on- and off-duty behavior in a foreign country.

3 On average, only 6 percent of participants selected "Don't know" for any individual item. For any individual scale, between 6 percent and 11 percent of participants indicated "Don't know" for one or more items.

4 Coefficient alpha is a measure of a scale's internal consistency (Nunnally and Bernstein, 1994). It is one of the methods used to determine how well items in a scale are working as a measure of the same psychological construct (i.e., the same concept). An increase in coefficient alpha when an item is removed from the scale suggests that the item does not measure the same construct as the other items in the scale. Common practice is to remove such an item from the scale.

5 Results for the shortened scale are labeled "SHORT" in all tables and figures. The items excluded in the shortened scale are noted in Appendix C.

of 1 to 5, with 1 = strongly disagree and 5 = strongly agree. Participants also had the option of selecting "USAF didn't train me to do this" rather than using the 1–5 scale.

Language skills. This section presented participants with two items in order to assess their language skills. The first item asked participants to rate their skill in any language other than English; the second asked them to rate their skill in the language of the country to which they had most recently deployed. The response options were based on the Interagency Language Roundtable (ILR) scale ranging from 0 to 5, with 0 = no language skills and 5 = complete fluency. (Appendix D provides weighted correlations and descriptive statistics for the two language self-assessments.)

Comprehensiveness of the 14 behavior categories. The last section of the survey solicited participants' input on the adequacy of our 14 categories for describing the domain of cross-cultural behavior of airmen. Participants were asked to respond to the following open-ended question: "The purpose of this survey is to establish the cross-cultural skills needed by Air Force personnel when deployed to foreign countries. We have identified the following 14 categories of such skills. If there is a skill that you believe is important and not represented in our list, please describe and explain in the comment box below." A list of our 14 behavior categories followed, along with a comment box.

Weighting Procedure

To ensure results were representative of the recently deployed Air Force population, we applied sample weights to some of the analyses, as noted in the discussion. Weights were computed as the ratio of the subpopulation size to the number of respondents from that subpopulation (Cochran, 1977). To compute the weights, we used population, sampling, and response rate information for each of our stratified subgroups.

As noted previously, the stratified subgroups were defined by the following demographics:

- two-digit AFSC
- grade: low (E-1 through E-4 for enlisted; O-1 through O-3 for officers) versus high (E-5 through E-9 for enlisted; O-4 through O-9 for officers)
- deployment location: Iraq/Afghanistan versus other.

Current (rather than during-deployment) AFSC and grade were used to estimate population sizes and therefore were also used to compute weights. Individuals who provided information only at the family level of AFSC were weighted to the AFSC family rather than to the specific two-digit AFSC. Individuals who did not provide sufficient information to be weighted (i.e., did not answer key demographic questions or indicated that their deployment location was classified) were assigned a weight of one, as were individuals whose weight would otherwise have been less than one.

Results

Importance of the Categories

Figure 3.1 shows the survey results on the importance of the 14 behavior categories, along with the results for a single item related to the foreign language skills category (discussed

Figure 3.1
Perceived Importance of Categories of Behavior

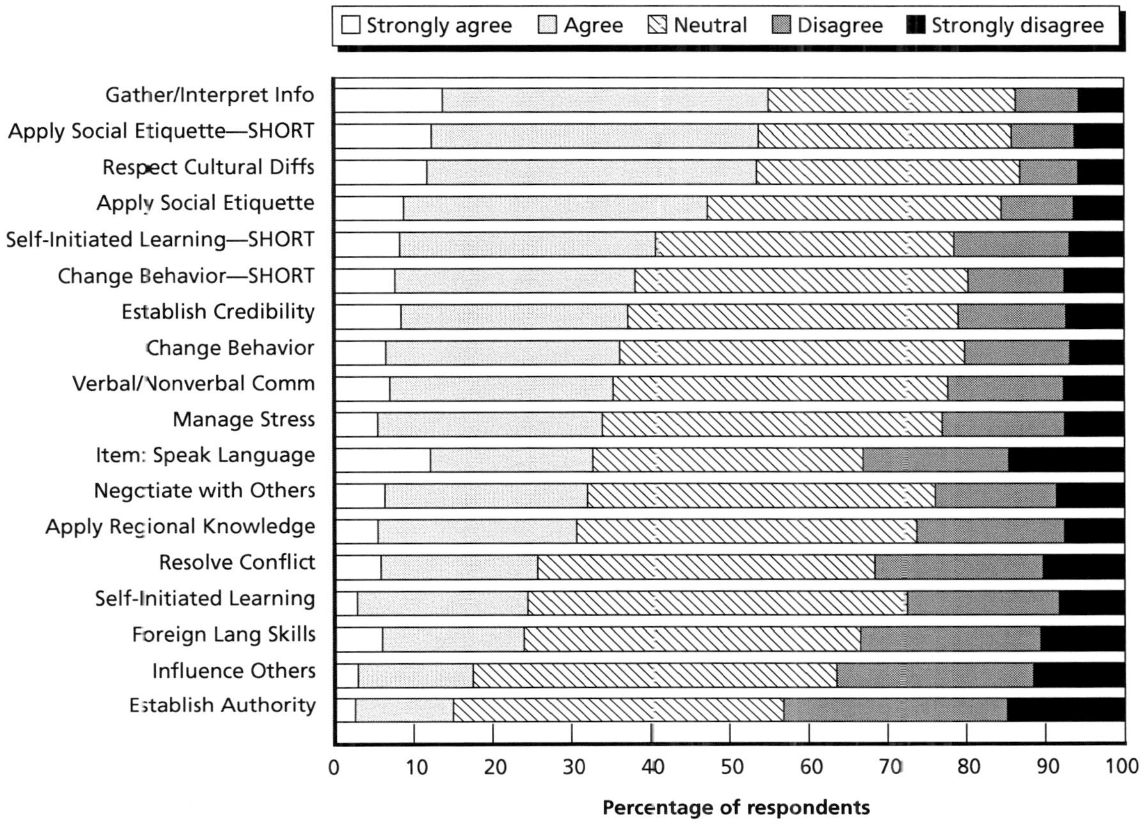

NOTES: (1) Responses for each scale were grouped as follows: strongly agree = 5.00 to 4.50, agree = 4.49 to 3.50, neutral = 3.49 to 2.50, disagree = 2.49 to 1.50, and strongly disagree = 1.49 to 1.00. (2) "SHORT" indicates that a scale was computed after removal of an item in the indicated category; this was done for the three categories for which removal of an item improved the scale's internal consistency.

RAND *MG811-3.1*

below).[6] All of the results are shown as proportions (weighted to represent the deployed population) of participants who strongly agree, agree, neither agree nor disagree, disagree, or strongly disagree that a behavior category is important. The categories are listed in order of importance—i.e., they run from the one with the highest proportion of persons agreeing (those who strongly agree + those who agree) that it is important (which in this case is gathering and interpreting observed information) to the one with the lowest proportion of such persons (establishing authority).

Several findings shown in Figure 3.1 are noteworthy. First, the foreign language skills category is low (third from the bottom) in order of importance—a result we found surprising given the Air Force's willingness to dedicate resources to developing airmen's language skills (e.g., the Air Force currently provides bonuses to airmen for having certain language skills). To further examine these results, we computed the proportion of responses to a survey item in this category that asked solely about speaking a language, since speaking may be more important than the other behaviors in this category. The results for this item alone (shown as "item: speak

[6] Note that, as discussed earlier, we show results for both the original scale and the shortened one.

language") were higher than for the category as a whole, suggesting that speaking is more important. However, this item was still ranked lower in importance than at least half of the remaining categories of cross-cultural behavior. While this does indicate moderate importance for the skill of speaking the language, it does not support the idea that foreign language skills training efforts should outrank training efforts in all other categories.

Second, the category of managing stress in an unfamiliar cultural setting shows up in a position of moderate importance. This suggests that deployed airmen think that managing stress in unfamiliar circumstances is an important skill to have. It also suggests that it may be important to include stress management in training that prepares airmen to work in foreign environments. This rating of stress management is also surprising in relationship to the Air Force's willingness to dedicate resources to improving this skill, but, in this case, surprising in terms of how high it is. This category was rejected by most training personnel in our informal interviews as the least relevant aspect of performance and was denounced as a skill that should not be included in Air Force training.

Third, all of the goal-oriented behavior categories—establishing authority, influencing others, resolving conflict, and negotiating with others—are in the bottom half of the importance ranking. This is consistent with our expectation that goal-oriented behaviors are highly specialized and therefore more likely to be needed only by airmen in certain AFSCs. In keeping with this theory, establishing authority, a type of behavior likely needed only by airmen working in a law enforcement role, had the lowest proportion of participants rating it important of any category.[7]

Last, for every category in Figure 3.1, at least some airmen indicated it was important. Given that the cut points we used to group scale scores into the "agree" and "strongly agree" categories in this figure were conservative, this finding suggests that no category should be rejected from inclusion in the final list of behavior categories for Air Force cross-cultural performance training.

Importance of the Categories by AFSC, Grade, and Deployment Location

We next examined the airmen's ratings of the importance of the behavior categories by subgroups—AFSCs, grades (enlisted and officer), and deployment locations. Table 3.1 shows what we found for 10 of the AFSCs (see Tables E.1 and E.2, in Appendix E, for results on all of the AFSCs). Several of our findings are noteworthy.

First, many AFSCs differ noticeably from others in their ratings of how important the categories of behavior are to their deployed performance. For example, as can be seen in the table, Pilots typically did not consider many behavior categories to be of even mild importance. Other AFSCs notable for their low importance ratings were Navigators (12), Aircrew Operations (1a), Weather (1w), some enlisted Logistics AFSCs (2a, 2e, 2g, 2p, 2w), Communications and Computers (3c), Mental Health Services and Medical Lab (4C and 4T), and Scientific/Research (61). In contrast, several AFSCs assigned relatively high importance to behavior categories in terms of both statistical significance and practical significance.[8] For example, Table 3.1 shows that regardless of grade or deployment location, Security Forces (31, 3P), Contract-

[7] It was, however, rated important by some personnel. A discussion about which AFSCs found this category important is included in the next subsection.

[8] AFSCs that are significant are those whose means are significantly higher than the means of the comparison group, Pilots.

Table 3.1
Average Importance Ratings of Behavior Categories by 10 AFSCs, by Grade and Deployment Location

Light Blue = Mild Importance Ratings of 3.1–3.4 | Medium Blue = Medium Importance Ratings of 3.5–3.9 | Dark Blue = High Importance Ratings of 4.0–5.0

2-Digit AFSC	Deployed Location	Grade	Gather/Interpret Info	Apply Social Etiquette—SHORT	Respect Cultural Diffs	Apply Social Etiquette	Self-Initiated Learning—SHORT	Change Behavior—SHORT	Establish Credibility	Change Behavior	Verbal/Nonverbal Comm	Manage Stress	Item: Speak Language	Negotiate with Others	Apply Regional Knowledge	Resolve Conflict	Self-Initiated Learning	Foreign Lang Skills	Influence Others	Establish Authority	# Respondents	Invited Sample	Population
All Deployed Personnel	**All Deployed Personnel**	**All Deployed Personnel**	**3.4**	**3.3**	**3.4**	**3.3**	**3.1**	**3.1**	**3.1**	**3.0**	**3.1**	**3.1**	**2.9**	**3.0**	**2.9**	**2.8**	**2.9**	**2.7**	**2.7**	**2.5**	**5,374-6,015**	**23,020**	**88,304**
Pilot (11XX)	**IRAQ/AFG**	LO	2.8	2.7	2.9	2.7	2.7	2.4	2.6	2.4	2.5	2.6	2.5	2.4	2.8	2.2	2.6	2.5	2.3	2.1	**26-33**	150	700
		HI	3.0	3.0	3.0	3.0	2.9	2.8	2.9	2.7	2.9	2.8	2.9	2.8	3.1	2.6	2.7	2.6	2.7	2.2	**57-68**	150	392
	OTHER	LO	3.0	3.0	3.0	3.0	2.7	2.7	2.7	2.6	2.6	2.5	2.7	2.7	2.7	2.4	2.6	2.5	2.2	2.1	**24-31**	150	2,711
		HI	3.0	3.2	3.3	3.2	2.8	2.9	2.7	2.8	2.8	2.7	2.9	2.7	2.8	2.3	2.5	2.6	2.3	1.9	**47-60**	150	962
Intelligence (14XX)	**IRAQ/AFG**	LO	3.6	3.6	3.6	3.6	3.6	3.4	3.3	3.3	3.2	3.3	3.5	3.0	4.1	2.9	3.4	3.3	2.8	2.5	**45-52**	150	236
		HI	3.8	3.9	3.9	3.9	3.5	3.7	3.4	3.6	3.7	3.3	3.5	3.4	3.9	3.1	3.3	3.3	3.3	2.7	**21-26**	68	68
	OTHER	LO	3.3	3.3	3.5	3.3	3.0	3.0	2.8	2.9	2.7	2.9	2.6	2.8	3.0	2.3	2.7	2.5	2.2	1.9	**38-46**	150	427
		HI	3.3	3.5	3.5	3.4	3.1	3.3	2.9	3.1	2.9	3.2	3.2	3.0	3.1	2.5	2.9	2.9	2.4	2.0	**35-36**	108	108
Intelligence (1NXXX)	**IRAQ/AFG**	LO	3.8	4.0	3.8	3.9	3.5	3.7	3.6	3.6	3.6	3.4	3.4	3.5	3.4	3.6	3.3	3.0	3.4	3.0	**11-15**	150	290
		HI	3.4	3.4	3.4	3.3	3.4	3.3	3.1	3.1	3.3	3.3	3.2	3.0	3.5	2.8	3.1	3.1	2.9	2.5	**49-54**	150	313
	OTHER	LO	3.3	3.5	3.5	3.5	3.2	3.3	2.9	3.1	3.2	3.1	2.5	3.1	3.2	2.4	3.0	2.4	2.4	2.1	**17-24**	150	436
		HI	3.0	3.0	3.7	3.7	3.4	3.6	3.2	3.4	3.3	3.2	3.1	3.2	3.2	2.9	3.2	3.0	2.7	2.3	**43-48**	150	437
Security Forces (31XX)	**IRAQ/AFG**	LO	4.4	4.2	4.2	4.2	3.9	4.0	4.2	3.9	4.3	3.9	4.1	4.0	4.0	4.0	3.6	3.7	4.0	3.8	**29-31**	150	178
		HI	4.5	4.4	4.3	4.4	4.1	4.2	4.4	4.1	4.3	3.7	4.4	4.1	4.0	4.1	3.8	3.8	4.2	3.9	**12**	45	45
	OTHER	LO	4.4	4.3	4.1	4.2	4.1	4.1	4.2	4.0	4.2	3.9	3.8	4.1	3.9	4.0	3.9	3.4	3.6	3.1	**7-9**	50	50
		HI																			**2**	12	12
Security Forces (Military Police) (3PXXX)	**IRAQ/AFG**	LO	4.0	3.8	3.7	3.7	3.2	3.2	3.3	3.1	3.7	3.7	3.2	3.3	2.9	3.8	2.9	2.8	3.0	3.6	**10-12**	150	3,442
		HI	4.0	3.8	3.7	3.7	3.5	3.6	3.7	3.4	3.9	3.6	3.7	3.5	3.5	3.6	3.3	3.3	3.4	3.5	**34-42**	150	2,206
	OTHER	LO	4.0	3.8	3.9	3.7	3.8	3.4	3.4	3.3	3.6	3.6	3.3	3.2	3.1	3.2	3.5	3.0	3.1	3.4	**11-15**	150	2,508
		HI	3.8	3.8	3.8	3.7	3.5	3.6	3.5	3.5	3.5	3.6	3.5	3.4	3.4	3.4	3.2	3.2	3.1	3.1	**37-41**	150	1,633
Contracting (64XX)	**IRAQ/AFG**	LO	4.2	4.3	4.2	4.2	3.9	4.0	4.1	3.8	4.0	3.8	3.7	4.3	3.7	3.5	3.6	3.2	3.4	3.0	**31-34**	111	111
		HI	3.8	4.1	4.1	4.1	3.8	3.5	3.9	3.5	3.8	3.3	3.6	4.2	3.8	3.4	3.4	3.2	3.5	3.0	**19-21**	52	52
	OTHER	LO	4.0	4.3	4.3	4.3	4.0	3.9	3.9	3.7	4.0	3.5	4.0	4.5	3.4	3.1	3.6	3.5	3.2	2.6	**20-22**	69	69
		HI																			**3-4**	4	4
Contracting (6CXXX)	**IRAQ/AFG**	LO	4.1	4.1	4.2	4.1	4.0	3.7	4.0	3.7	3.9	3.9	4.1	4.2	3.7	3.6	3.6	3.3	3.3	3.3	**7-8**	51	51
		HI	3.9	4.1	4.1	4.0	3.7	3.7	3.9	3.6	3.9	3.7	3.3	4.2	3.3	3.3	3.5	2.9	3.3	3.0	**53-60**	150	211
	OTHER	LO	4.1	4.4	4.3	4.3	3.9	3.8	3.9	3.5	3.9	3.4	3.4	4.2	3.6	2.9	3.6	3.0	3.3	3.1	**6-8**	59	59
		HI	3.6	3.8	3.8	3.8	3.4	3.5	3.6	3.3	3.6	3.2	3.0	4.0	3.1	3.1	3.2	2.9	3.0	2.7	**46-51**	150	168
International Affairs/FAO (16XX)	**IRAQ/AFG**	LO	4.4	4.2	4.3	4.1	4.3	4.1	4.3	4.0	4.1	4.1	4.5	4.0	4.2	4.0	3.9	4.0	3.8	3.2	**10-13**	6	6
		HI	4.0	4.0	3.9	4.0	3.9	3.8	4.0	3.7	3.9	3.6	4.0	3.7	3.8	3.4	3.5	3.4	3.7	2.8	**45-48**	39	39
	OTHER	LO	3.6	3.5	3.2	4.0	3.4	3.3	2.7	3.1	2.3	2.7	2.3	2.5	3.0	2.3	2.9	1.9	2.2	1.7	**5-6**	5	5
		HI	3.6	3.9	3.8	3.7	3.6	3.5	3.3	3.4	3.4	3.3	3.4	3.4	3.0	2.6	3.3	3.3	2.7	1.9	**17-21**	43	43

RAND *MG811-T.3.1a*

Table 3.1—continued

2-Digit AFSC	Deployed Location	Grade	Gather/Interpret Info	Apply Social Etiquette—SHORT	Respect Cultural Diffs	Apply Social Etiquette	Self-Initiated Learning—SHORT	Change Behavior—SHORT	Establish Credibility	Change Behavior	Verbal/Nonverbal Comm	Manage Stress	Item: Speak Language	Negotiate with Others	Apply Regional Knowledge	Resolve Conflict	Self-Initiated Learning	Foreign Lang Skills	Influence Others	Establish Authority	# Respondents	Invited Sample	Population
Special Investigations (OSI) (7SXXX)	IRAQ/AFG	LO																			**0**	0	0
		HI	4.4	4.6	4.2	4.5	3.9	4.2	4.3	4.1	4.4	3.9	4.2	4.3	4.2	3.7	3.6	3.8	3.9	3.4	**23-25**	121	121
	OTHER	LO																			**0**	0	0
		HI	4.4	4.5	4.2	4.5	4.4	4.3	4.2	4.2	4.3	4.2	3.8	4.3	4.2	3.7	4.0	3.8	4.0	2.8	**7-8**	81	81
TCN Escort	IRAQ/AFG	LO	3.7	3.3	3.4	3.2	3.1	3.0	3.2	2.9	3.3	3.3	3.2	2.9	3.0	3.2	2.8	2.8	3.1	3.3	**22-27**	0	0
		HI	4.1	3.6	3.9	3.6	3.6	3.2	3.7	3.0	3.8	3.5	3.6	3.2	3.1	3.5	3.3	3.1	3.4	3.6	**20-23**	0	0
	OTHER	LO	3.9	3.9	3.8	3.8	3.4	3.5	3.5	3.4	3.7	3.5	3.4	3.3	3.2	3.3	3.3	2.9	3.2	3.2	**35-43**	0	0
		HI	3.9	3.9	3.9	3.9	3.6	3.5	3.6	3.5	3.8	3.4	3.3	3.3	3.2	3.2	3.3	2.8	3.3	3.3	**22-28**	0	0

NOTES: (1) The weighted average for all deployed personnel is in the top row, for comparison. Except for the data in that row, all results are unweighted. The estimates in the rows are representative of the ratings of those who deployed in that particular AFSC, grade, and location. Because this level of detail is precise enough to enable near-identification of participants (to whom we guaranteed confidentiality of responses) and because results based on few respondents are very unreliable, no results are shown in rows where the number of survey respondents was less than five. (2) Importance ratings within the table are color-coded such that light blue represents "mild" importance ratings (ranging from 3.1 to 3.4), medium blue represents "medium" importance ratings (ranging from 3.5 to 3.9), and dark blue represents "high" importance ratings (ranging from 4.0 to 5.0). These divisions were chosen such that all importance ratings that are stronger than neutral (a rating of 3.00) are highlighted in blue, with the darkest blue representing the strongest level of agreement.

RAND *MG811-T.3.1b*

ing (64, 6C), Special Investigations OSI (7S), and International Affairs (16) rated many of the 14 categories as high in importance. It also shows that respondents who wrote-in Third-Country National (TCN) Escort as their deployed AFSC tended to rate the categories as high. Table 3.2 shows the rank order of the AFSCs after their importance ratings were averaged across all categories (after controlling for grade, deployment location, and deployed mission).

Table 3.2
Rank Order of AFSCs by Importance Ratings Averaged Across All Categories

Rank	AFSC	Rank	AFSC
1	Special Investigations (OSI) (7SXXX)	31	Mission Support (3SXXX)
2	Security Forces (31XX)	31	Chaplain Assistant (5RXXX)
3	Special Investigations (OSI) (71XX)	32	Biomedical Specialists (43XX)
4	Support Commander (30XX)	32	Law (51XX)
5	Contracting (64XX)	33	Maintenance Management Systems (2RXXX)
6	Contracting (6CXXX)	34	Safety (1SXXX)
7	Public Affairs (3NXXX)	34	Information Management (3AXXX)
7	Security Forces (Military Police) (3PXXX)	34	Civil Engineering (3EXXX)
8	Surgery (45XX)	34	Acquisition (63XX)
9	Unallotted Airman (9UXXX)	35	Aerospace Medicine (48XX)
10	Public Affairs (35XX)	35	Medical Laboratory (4TXXX)
11	Civil Engineer (32XX)	36	Intelligence (14XX)
12	Cardiopulmonary Laboratory (4HXXX)	37	Weather (15XX)
13	TCN Escort	38	Health Services Administrator (41XX)
14	Chaplain (52XX)	38	Biomedical Clinician (42XX)
15	International Affairs/FAO (16XX)	39	Command & Control Systems Ops (1CXXX)
16	Visual Information (3VXXX)	39	Transportation & Vehicle Maintenance (2TXXX)
17	Aircrew Protection (1TXXX)	40	Paralegal (5JXXX)
18	Services (34XX)	41	Aircraft Maintenance (21XX)
19	Services (3MXXX)	42	Aircrew Operations (1AXXX)
19	Postal Specialist (8MXXX)	42	Scientific/Research (61XX)
20	No AFSC Provided	43	Precision Measurement Equipment Lab (2PXXX)
21	Intelligence (1NXXX)	43	Comm-Information Systems (33XX)
22	Financial (6FXXX)	44	Munitions & Weapons (2WXXX)
23	Nurse (46XX)	45	Manned Aerospace Maintenance (2AXXX)
23	Physician (44XX)	45	Communications & Computer Systems (3CXXX)
24	Group Superintendent (9GXXX)	46	Communications & Electronics (2EXXX)
25	Space, Missile, and Command & Control (13XX)	47	Navigator (12XX)
26	Fuels (2FXXX)	48	Mental Health Services (4CXXX)
26	Finance (65XX)	49	Logistics Plans (2GXXX)
27	Medical Service (4NXXX)	50	Weather (1WXXX)
28	Medical (4AXXX)	51	Pilot (11)
29	Manpower-Personnel (37XX)		
30	Supply (2SXXX)		
30	First Sergeant (8FXXX)		

NOTES: (1) These are the results after controlling for grade, deployment location, and deployed mission. For full regression results, see Appendix F. (2) Some rankings are shared.

Second, after controlling for AFSC differences and deployed mission, we found that deployment location (Iraq/Afghanistan versus Other) and grade (low versus high) did not have a significant effect on the importance ratings averaged across the categories (regression results predicting the importance ratings averaged across the categories are in Appendix F, Table F.1). However, we did observe some within-AFSC differences based on grade in a few AFSCs. For example, as shown for Intelligence (1N) in Table 3.1, the low-grade group that deployed to Iraq/Afghanistan rated more categories as important than did the high-grade group. Similarly, we found some differences by deployment location in a few AFSCs. For example, for Intelligence (14), we observed higher ratings of importance for those who deployed to Iraq/Afghanistan than for those who deployed to other locations. However, there were no visibly or statistically consistent findings for grade differences or deployment location differences across all AFSCs.

In addition, as shown in Table 3.3, we found significant differences by grade (low, as compared with high) and deployment location within certain categories of behavior. When we controlled for AFSC, deployment location, and deployed mission, the categories of managing stress in an unfamiliar cultural setting and establishing authority were seen as more important by respondents at lower grades than by respondents at higher grades. Differences for deployment location when we controlled for grade, AFSC, and deployed mission presented a more complicated picture. Respondents deployed to Iraq/Afghanistan perceived the establishment of authority as more important than did respondents deployed to other locations, but they also perceived several categories as less important than did those deployed to other locations.

Third, in general it appears that the importance an AFSC assigned to one category of behavior was likely to carry over to the importance assigned to other categories. Pilots, for

Table 3.3
The Effect of Grade and Deployment Location on Importance Ratings, Summary of Findings

Category	Low Grade	Iraq/Afghanistan
Overall	NS	NS
Gathering and Interpreting Observed Information	Significantly higher	Significantly lower
Applying Appropriate Social Etiquette—SHORT	NS	Significantly lower
Respecting Cultural Differences	Significantly higher	Significantly lower
Applying Appropriate Social Etiquette	Significantly higher	Significantly lower
Self-Initiated Learning—SHORT	NS	NS
Changing Behavior to Fit Cultural Context—SHORT	NS	Significantly lower
Establishing Credibility, Trust, and Respect	NS	NS
Changing Behavior to Fit Cultural Context	NS	Significantly lower
Verbal and Nonverbal Communication	NS	Significantly lower
Managing Stress in an Unfamiliar Cultural Setting	Significantly higher	NS
Negotiating with Others	NS	Significantly lower
Applying Regional Knowledge	NS	NS
Resolving Conflict	NS	NS
Self-Initiated Learning	Significantly higher	Significantly lower
Foreign Language Skills	NS	NS
Influencing Others	NS	NS
Establishing Authority	Significantly higher	Significantly higher

NOTES: (1) These are the results after controlling for AFSC and deployed mission. For full regression results, see Appendix F. (2) NS = not significantly related to ratings of importance. (3) Low grade is in comparison to high grade; Iraq/Afghanistan is in comparison with all other deployment locations.

example (see Table 3.1) tended to assign low ratings across all categories, and Security Forces tended to assign high ratings to nearly every category. However, we did find exceptions to this pattern.

One exception concerns Contracting (see Table 3.1), for which the ratings for the category of negotiating with others are noticeably higher than those for many other categories. As further illustration, consider the results shown in Table 3.4, which lists, for each of the 14 behavior categories, the top 10 AFSCs in terms of highest importance ratings. Each group of 10 is arranged according to perceived importance of the category, from highest to lowest; as can be seen, the order within the groups is not the same for all categories. For example, when deployment location, grade, and deployed mission were controlled for, both enlisted and officer OSI (7S and 71) respondents consistently rated the categories significantly higher than did respondents in other AFSCs, as might be expected. The only exception occurs for foreign language skills, for which the importance ratings of OSI officers were lower than those of individuals in several other AFSCs, such as Security Forces (31), Support Commanders (30), Public Affairs (35), and enlisted OSI (7S). The AFSCs that rated nine or more of the 14 behaviors consistently and significantly as more important than did other AFSCs were Security Forces (3P, 31), Public Affairs (3N, 35), Contracting (6C, 64), and Support Commanders (30). Foreign Area Officers (FAOs) rated many of the categories significantly higher than did the other AFSCs; however, they only reached the top 10 highest ratings for the category of applying regional knowledge. Lastly, 173 people selected Other as their deployed AFSC and wrote-in TCN Escort. This group rated verbal and nonverbal communication, influencing others, and establishing authority as significantly more important than did many other AFSCs.

The AFSCs highlighted in this discussion were not the only AFSCs that tended to rate the behavior categories highly, but they were the AFSCs that produced the strongest regression results (i.e., their regression coefficients had the largest magnitudes). Many other AFSCs rated various categories as significantly more important than did Pilots. However, given our large sample size, we considered both practical significance (i.e., effect size) and statistical significance in choosing which results to present. To examine other AFSC differences, see Appendix E (individual AFSC results) and Appendix F (regression results).

Perceptions of Training

Proportion Reporting Having Received Training. Overall, the proportion of respondents reporting that they had received training was small. Figure 3.2 shows the weighted percentage of respondents who reported being trained just prior to deployment and over the course of their career for each of the 14 behavior categories. As can be seen, the largest proportions of training-received reports (which occurred for such categories as respecting cultural differences and self-initiated learning) were around 75 percent, which means that even in these cases, 25 percent of respondents indicated that they had *not* received training. For some categories, less than 60 percent reported having been trained. Note that the foreign language skills category had the lowest proportion of all: More than 50 percent of respondents indicated that they had *not* received training either just prior to deployment or over the course of their career.

For nearly all of the behavior categories, the reported rates for over-career training were only slightly higher than those for predeployment training. This is surprising, since one would expect there to be decidedly more training over the course of a career than just before deployment. It is possible, however, that the survey participants were more likely to recall their predeployment training experiences because of their recency.

Table 3.4
AFSCs Assigning Top 10 Highest Ratings to the Behavior Categories

Gather/Interpret Information	Apply Social Etiquette—SHORT	Respect Cultural Diffs
1. Special Investigations (OSI) (7S)	1. Special Investigations (OSI) (7S)	1. Contracting (64)
2. Security Forces (31)	2. Special Investigations (OSI) (71)	2. Support Commander (30)
3. Support Commander (30)	3. Contracting (64)	3. Special Investigations (OSI) (7S)
4. Contracting (64)	4. Support Commander (30)	4. Surgery (45)
5. Special Investigations (OSI) (71)	5. Security Forces (31)	5. Physician (44)
6. Unallotted Airman (9U)	6. Contracting (6C)	6. Contracting (6C)
7. Public Affairs (3N)	7. Public Affairs (35)	7. Security Forces (31)
8. Civil Engineer (32)	8. Public Affairs (3N)	8. Public Affairs (3N)
9. Contracting (6C)	9. Services (34)	9. Special Investigations (OSI) (71)
10. Surgery (45)	10. Surgery (45)	10. Public Affairs (35)

Apply Social Etiquette	Self-Initiated Learning—SHORT	Change Behavior—SHORT
1. Special Investigations (OSI) (7S)	1. Special Investigations (OSI) (7S)	1. Special Investigations (OSI) (7S)
2. Special Investigations (OSI) (71)	2. Special Investigations (OSI) (71)	2. Special Investigations (OSI) (71)
3. Contracting (64)	3. Contracting (64)	3. Security Forces (31)
4. Support Commander (30)	4. Support Commander (30)	4. Contracting (64)
5. Contracting (6C)	5. Security Forces (31)	5. Support Commander (30)
6. Security Forces (31)	6. Contracting (6C)	6. Public Affairs (3N)
7. Public Affairs (3N)	7. Public Affairs (3N)	7. Contracting (6C)
8. Services (34)	8. Surgery (45)	8. Public Affairs (35)
9. Surgery (45)	9. Chaplain (52)	9. Cardiopulmonary Lab (4H)
10. Public Affairs (35)	10. Public Affairs (35)	10. Surgery (45)
	10. Security Forces (3P)	

Establish Credibility	Change Behavior	Verbal/Nonverbal Comm
1. Special Investigations (OSI) (7S)	1. Special Investigations (OSI) (7S)	1. Special Investigations (OSI) (7S)
2. Support Commander (30)	2. Special Investigations (OSI) (71)	2. Special Investigations (OSI) (71)
3. Contracting (64)	3. Support Commander (30)	3. Security Forces (31)
4. Special Investigations (OSI) (71)	4. Contracting (64)	4. Contracting (64)
5. Contracting (6C)	5. Security Forces (31)	5. Support Commander (30)
6. Surgery (45)	6. Public Affairs (35)	6. Contracting (6C)
7. Security Forces (31)	7. Contracting (6C)	7. Civil Engineer (32)
8. Public Affairs (3N)	8. Public Affairs (3N)	8. Unallotted Airman (9U)
9. Civil Engineer (32)	9. Cardiopulmonary Lab (4H)	9. Surgery (45)
10. Public Affairs (35)	10. Chaplain (52)	10. Public Affairs (35)
10. Security Forces (3P)	10. Security Forces (3P)	

Table 3.4—continued

Manage Stress	Negotiate with Others	Apply Regional Knowledge
1. Special Investigations (OSI) (71)	1. Contracting (64)	1. Special Investigations (OSI) (7S)
2. Special Investigations (OSI) (7S)	2. Special Investigations (OSI) (7S)	2. Security Forces (31)
3.Security Forces (31)	3. Contracting (6C)	3. Support Commander (30)
4. Cardiopulmonary Lab (4H)	4. Special Investigations (OSI) (71)	4. Contracting (64)
5. Support Commander (30)	5. Security Forces (31)	5. Special Investigations (OSI) (71)
6. Surgery (45)	6. Support Commander (30)	6. Public Affairs (35)
7. Public Affairs (3N)	7. Civil Engineer (32)	7. Public Affairs (3N)
8. Contracting (64)	8. Public Affairs (3N)	8. Chaplain (52)
9. Contracting (6C)	9. Public Affairs (35)	9. Intelligence (14)
10. Security Forces (3P)	10. Security Forces (3P)	10. International Affairs/FAO (16)
	10. Services (34)	10. Security Forces (3P)

Resolve Conflict	Self-Initiated Learning	Foreign Lang Skills
1. Security Forces (31)	1. Special Investigations (OSI) (7S)	1. Special Investigations (OSI) (7S)
2. Special Investigations (OSI) (7S)	2. Contracting (64)	2. Security Forces (31)
3. Special Investigations (OSI) (71)	3. Security Forces (31)	3. Support Commander (30)
4. Support Commander (30)	4. Special Investigations (OSI) (71)	4. Public Affairs (35)
5. Security Forces (3P)	5. Support Commander (30)	5. Contracting (64)
6. Contracting (64)	6. Contracting (6C)	6. Postal Specialist (8M)
7. Unallotted Airman (9U)	7. Surgery (45)	7. Security Forces (3P)
8. Civil Engineer (32)	8. Public Affairs (3N)	8. Surgery (45)
9. Public Affairs (35)	9. Security Forces (3P)	9. Chaplain (52)
10. Contracting (6C)	10. Chaplain (52)	10. Public Affairs (3N)
10. TCN Escort		

Influence Others	Establish Authority
1. Special Investigations (OSI) (7S)	1. Security Forces (31)
2. Security Forces (31)	2. Unallotted Airman (9U)
3. Special Investigations (OSI) (71)	3. Security Forces (3P)
4. Support Commander (30)	4. TCN Escort
5. Contracting (64)	5. Special Investigations (OSI) (71)
6. Unallotted Airman (9U)	6. Special Investigations (OSI) (7S)
7. TCN Escort	7. Contracting (6C)
8. Security Forces (3P)	8. Civil Engineer (32)
9. Contracting (6C)	9. Support Commander (30)
10. Surgery (45)	10. Cardiopulmonary Lab (4H)

NOTES: (1) These are the results after controlling for grade, deployment location, and deployed mission. For full regression results, see Appendix F. (2) In some cases, as shown, two AFSCs tied for the 10th highest rating.

Figure 3.2
Training Received in Behavior Categories

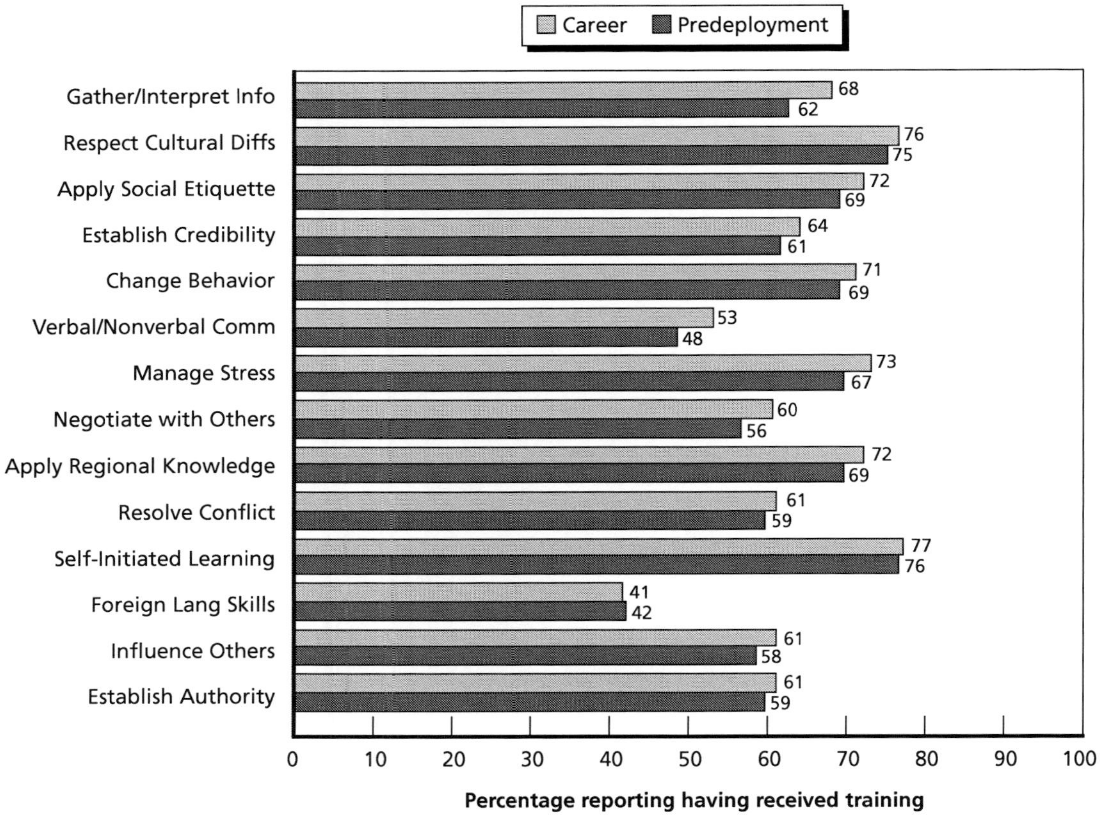

NOTE: For easier comparison across figures, categories are ordered consistently with the importance rankings in Figure 3.1.
RAND *MG811-3.2*

Ratings of the Helpfulness of Training. Ratings of training's helpfulness were provided only by those who indicated that they had received training. Figure 3.3 shows the percentages of these respondents agreeing or strongly agreeing that the training they received in the 14 categories was helpful. As can be seen, relatively few thought their training was helpful. This was particularly true for the categories of foreign language skills and verbal and nonverbal communicating. Training in the category of respecting cultural differences was regarded as the most helpful. For all categories, over-career training was rated more helpful than was predeployment training.

Perceptions of Training by AFSC, Grade, and Deployment Location

Proportion Reporting Having Received Training. We conducted regression analyses for both predeployment and over-career training to test for significant differences in the training received across AFSCs, grades, and deployment locations.

A few of the regression analysis results for predeployment training were surprising—for example, FAOs reported receiving this form of training in significantly fewer categories than did Pilots. Unsurprisingly, OSI (71, 7S) indicated that they received predeployment training in significantly more of the 14 categories than did Pilots. Other AFSCs that reported

Figure 3.3
Helpfulness of Training in Behavior Categories

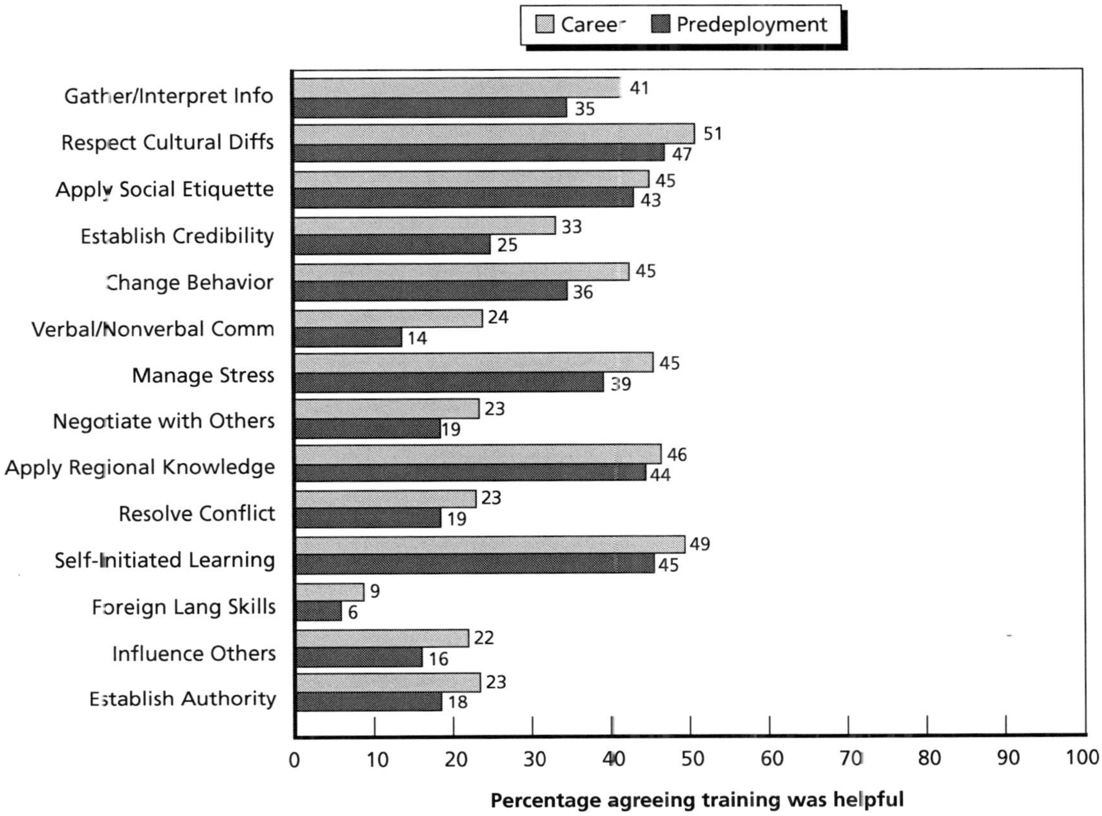

NOTE: For easier comparison across figures, categories are ordered consistently with the importance rankings in Figure 3.1.
RAND *MG811-3.3*

receiving predeployment training in significantly fewer categories than did pilots were Safety (1S), Aircraft Maintenance (21), and Comm-Info Systems (33). Grade also made a difference: Lower grades reported having received predeployment training in significantly more categories than did higher grades. Deployment location, however, made no difference.

The regression analysis results for over-career training were slightly different. Compared with Pilots, FAOs again reported receiving training in significantly fewer categories. Other AFSCs that reported receiving significantly fewer categories of training over the course of their career compared with Pilots were Safety (1S) and Precision Measurement Equipment (2P). Support commanders (30) and OSI (71, 7S) reported receiving over-career training in more categories than did Pilots.[9] Grade did not make a difference in predicting over-career training,

---

9 Regressions were weighted, and AFSCs used in the regressions were the deployed AFSCs. Since some people reported a current AFSC that differed from their deployed AFSC, the regressions described in the text include people whose deployed and current AFSCs were unmatched. We also ran a second set of regressions for over-career and predeployment training that included only those people whose current and deployed AFSCs matched. The results for this matched-AFSC set were consistent with the original results across both the over-career and the predeployment regressions except as follows: The FAO (16), Safety (1S), and Aircraft Maintenance (21) results were no longer significant, and the Biomedical Specialist (43) and Cardiopulmonary Lab (4H) results became significant.

but deployment location did: Those deployed to locations other than Iraq/Afghanistan were significantly more likely to have received training in more categories.

We also ran a separate regression to relate enlisted/officer, grade, and deployment location to the number of categories in which training was received. For both over-career and predeployment training, enlisted respondents reported receiving training in significantly more categories than did officers; lower grades reported receiving training in significantly more categories than did higher grades; and deployment location made no significant difference.

The surprising finding that FAOs reported being trained in significantly fewer categories than did Pilots may be the result of data limitations (discussed below). For example, it is very likely that the job requirements of FAOs demand more-stringent culture-training standards than do the job requirements of Pilots. These results therefore could suggest that FAOs perceive the training they are receiving as insufficient for meeting FAO standards, whereas Pilots perceive the training they are receiving as meeting Pilot standards.

Ratings of the Helpfulness of Training. We conducted a final set of regression analyses to examine whether deployment location, grade, and AFSC affected perceptions of the helpfulness of training.[10] For each respondent, a separate average helpfulness score was computed across all 14 categories for both predeployment training and over-career training. Because only the respondents forming the subset of those who reported receiving training were available to rate the helpfulness of training, the response sizes for many subgroups were small. Nevertheless, a few findings are noteworthy.

For the predeployment regression, the helpfulness ratings of most of the AFSC groups were significantly higher than those of Pilots. FAOs (16), Public Affairs (35, 3N), and TCN Escorts were exceptions, with ratings not significantly different from those of Pilots. Lower grades were significantly more likely than higher grades to indicate that predeployment training was helpful, regardless of AFSC and deployment location. Deployment location did not relate significantly to rated helpfulness of predeployment training. Results of a separate regression using enlisted/officer, deployment location, and grade as predictors showed that enlisted personnel were more likely than officers to rate predeployment training as helpful; lower grades, compared with higher grades, also found it more helpful. However, there were again no differences by deployment location for predeployment training.

As for the ratings of helpfulness of over-career training, AFSC significantly predicted helpfulness ratings. However, relative to the results for predeployment training, fewer AFSCs rated the training as significantly more helpful than did Pilots. Once more, FAOs (16) and Public Affairs (35, 3N) did not differ significantly from Pilots in their view of the helpfulness of the over-career training they had received, although groups such as OSI (71, 7S), Security Forces (31, 3P), and Contracting (6C) still did. Grade and deployment location did not relate significantly to helpfulness ratings. In a separate regression using enlisted/officer, deployment location, and grade, we found that none of these three could predict helpfulness ratings of over-career training.

---

[10] As with the regression analyses for reports of received training, we used weighting and deployed AFSC in our regression analyses for reports of helpfulness of training, which means we included people whose deployed and current AFSCs were unmatched. We thus decided to run a second regression for helpfulness that would include only those people whose current and deployed AFSCs matched. The differences in the AFSC-matched results were as follows: For predeployment training, Contracting (64) and First Sergeant (8F) were no longer significant, and grade became only marginally significant. For over-career training, Acquisition (63), First Sergeant (8F), and Postal Specialist (8M) were no longer significant, and Fuels (2F) became significant.

Additional results by AFSC, grade, and deployment location are provided in Appendix E: Tables E.3 through E.6 show the proportions of respondents reporting training in each of the 14 categories; Tables E.7 through E.10 show the weighted average levels of agreement that training was helpful in each of the 14 categories. These results on training can be used in conjunction with the training importance ratings—provided in Appendix E's Tables E.1 and E.2—to determine where to focus training resources in the short term. We suggest that training resources be directed first at those AFSCs that rated specific categories as highly or moderately important and also tended to rate existing training as not provided or not helpful. In addition, the training provided to members of AFSCs that reported their received training as helpful might be good sources of training that could be broadened to serve a more-general audience. We therefore suggest that these apparently AFSC-specific training programs rated as helpful be more closely examined, a subject we discuss further in Chapter Four.

Training Data Limitations. Our results clearly demonstrate that much of the current training is not meeting the perceived needs of the deployed Air Force population. However, two limitations of the training data are worth noting.

First, *training* was not defined in the survey, so standards for what constitutes training may have differed across respondents, categories of behavior, or both. Some respondents, for instance, may think a one-hour briefing constitutes training, whereas others may not. And some may consider a briefing to be helpful training for one category but not for another. Moreover, because the military sometimes distinguishes between training and education, it is possible that some participants would have responded differently had we asked about *training and education* rather than just training.

This limitation can make interpretation of the training results difficult. For example, very few respondents rated foreign language skills training as helpful. Also, less than 50 percent reported receiving language training. Respondents could have been considering foreign language skills training to be a brief introduction to common language phrases that they had received and thus concluding that training in this category is not helpful. However, if they were considering a few phrases provided in a briefing to be training, then the relatively low numbers of respondents reporting that they had received foreign language skills training would be surprising. Conversely, those who reported receiving language training may be largely those who have had extensive language training. If that is the case, then their ratings of that training as not helpful are noteworthy. The lack of usefulness of extensive language training was echoed in the survey's comments section by respondents' notes reporting that they had not deployed to the country in whose language they had trained (see section, below, on the comprehensiveness of the 14 categories).

Second, we only assessed the respondents' *perceptions* of training. We did not evaluate the form or depth of training curricula; neither did we measure actual improvement in performance. Rather, participants rated the existence and helpfulness of training for performance in the deployment location. Perceptions of training's helpfulness are often not correlated with actual training effectiveness (Alliger, Tannenbaum, Bennett, Traver, and Shotland, 1997). We therefore suggest that the training helpfulness ratings be viewed with caution. Additional suggestions for how to further evaluate training effectiveness are in Chapter Four.

Language Skills

Table 3.5 shows the weighted proportions of respondents at each level of language skill. The results show that only 19 percent of the respondents reported having any skill in the

Table 3.5
Prevalence of Language Skills

	Percentage Agreeing or Strongly Agreeing with Response When Asked About:	
Level of Proficiency	**Language of Deployment Location**	**Foreign Language I Know Best**
Do not speak, write, or understand the language	81	53
Satisfy routine travel needs and minimum courtesy requirements	14	28
Satisfy routine social demands and limited work requirements	3	9
Speak the language with sufficient structural accuracy and vocabulary to participate effectively in most formal and informal conversations on practical, social, and professional topics	1	5
Use the language fluently and accurately on all levels normally pertinent to professional needs	1	2
Speak with the proficiency equivalent to that of an educated native speaker	2	3
Total	100	100

NOTE: Rounding may prevent columns from adding exactly to 100 percent.

language of the country to which they deployed, and less than 4 percent considered themselves even modestly fluent.

Considering the respondents' low ratings of importance for the foreign language skills category (see earlier discussion), the finding that so few can speak the language of the deployment location was not surprising. Nevertheless, it led us to consider whether knowing how to speak a language might cause one to overestimate the importance of foreign language skills or, conversely, not knowing how to speak a language might cause one to underestimate its importance. Indeed, self-ratings of skill in the foreign language known best and the language of the deployment location are weakly but positively correlated with importance ratings for all 14 of the behavior categories (see Appendix D).

More importantly, the moderate relationship between foreign language skills and self-reported level of skill in the deployment language ($r = .22$) suggests that people may be unaware of the importance of language skills unless they actually can speak the language. This alone could explain why so few people agreed or strongly agreed that the foreign language skills category was important. Nevertheless, direction of causality cannot be determined solely from this observed correlation. An alternative explanation is that people who speak the language also hold jobs where speaking the deployment language is very important. To illustrate, Crypto-Linguists must comprehend the spoken language. They would therefore rate their language skills highly and would also indicate that the foreign language skills category is highly important to their job. In this case, the importance ratings would not be falsely inflated by language skills but, rather, would reflect an accurate evaluation of the importance of the job. Similarly, Pilots, who are probably much less likely to need to speak the language of the deployment country, would be unlikely to speak the language of the deployment country and to rate foreign language skills as important to their job.

To explore this possibility further, we compared airmen with AFSCs for which speaking the deployment language was particularly relevant or even a requirement for job performance with Pilots, for whom language skills would not typically be considered relevant. Our "language-relevant AFSCs" were FAOs (16), Crypto-Linguists (1N3) Airborne Cryptologic

Linguist (1A8X1), Security Forces (31PX), Public Affairs (35PX), OSI (7S), Linguist Debriefer (8D000), and Interpreter/Translator (9L000).

To ensure that we would be assessing only those whose deployment-location language was other than English, we removed individuals deployed to locations like Canada, Diego Garcia, and Cuba from the analyses. We also removed individuals whose deployed and current AFSCs did not match, which was often the case given that many individuals in our sample had either changed their duty AFSC since deployment or had deployed into an AFSC not consistent with their duty AFSC. We wanted to include only those individuals who typically held a language-relevant AFSC, not those who had been temporarily assigned to a language-relevant AFSC for deployment. Put another way, we attempted to have our language-relevant AFSC group consist only of people who had been well prepared for a language-relevant AFSC or had held such an AFSC for a long period.[11] We ended up with 163 individuals in the language-relevant AFSC group, which we compared with a group of 197 Pilots.

We then split these two groups by self-reported level of fluency in the foreign language they knew best and in the foreign language of their deployment location. Those who indicated they could do nothing were placed in the "no skill" group; those that self-reported any level of fluency (ranging from satisfying routine travel needs and minimum courtesy requirements to speaking the language with the fluency of an educated native speaker) were included in the "some skill" group. Table 3.6 displays the unweighted percentage of respondents with no versus some language skills in the two groups. Almost two-thirds of those with language-relevant AFSCs indicated that they did not speak the language of the deployment location with any fluency.

For the four subgroups—no skill/Pilots, some skill/Pilots, no skill/language-relevant AFSCs, and some skill/language-relevant AFSCs—we examined participants' average importance ratings for the foreign language skills category, the verbal and nonverbal communication category, and the single item "I found myself in situations (on and off duty) where it was or would have been important for me to speak the language of the country where I was deployed." Table 3.7 displays the results.

As shown in the table, the language-relevant AFSCs assigned more importance to all three language-related behavior measures than did Pilots. However, Pilots with no skill was the only subgroup whose importance ratings for deployment-location language differed sig-

Table 3.6
Comparison of Prevalence of Language Skills for Language-Relevant AFSCs and Pilots

	Percentages of Respondents Indicating Skill Level					
	Best-Known Language			Deployment Language		
	No Skill	Some Skill	Total	No Skill	Some Skill	Total
Language-relevant AFSC group	32	68	100	61	39	100
Pilots	67	33	100	88	12	100

[11] We acknowledge that our approach would exclude any individuals who had held a language-relevant AFSC for a long period but whose current AFSC did not reflect that job. However, for this particular analysis, we were most concerned that those included have extensive experience in a language-relevant AFSC (no false positives), not that someone excluded had such qualifications (false negatives).

Table 3.7
Comparison of Average Ratings of Importance of Language-Related Behaviors for Language-Relevant AFSCs and Pilots

	Average Importance Rating			
	Best-Known Language		Deployment Language	
Behavior Category	No Skill	Some Skill	No Skill	Some Skill
Language-relevant AFSC group				
Foreign Language Skills	3.51	3.56	3.50	3.63
Item: Speak the Language of Deployment[a]	4.02	3.80	3.76	4.11
Verbal/Nonverbal Communication	4.23	3.97	4.02	4.12
Pilots				
Foreign Language Skills	2.48	2.50	2.37	3.33
Item: Speak the Language of Deployment	2.56	2.72	2.52	3.79
Verbal/Nonverbal Communication	2.75	2.66	2.55	3.56

[a] Full wording of survey item: "I found myself in situations (on and off duty) where it was or would have been important for me to speak the language of the country where I was deployed."

nificantly from those of the three other subgroups. This outcome provides tentative support for the idea that possessing language skills might cause people to perceive language skills as more important. Further research is needed, however, to understand the nature of the relationship. Our analyses of the language-relevant AFSCs were exploratory and intended solely to begin looking at some of the issues raised by the results suggesting that the foreign language skills category is perceived as less important than several other categories of cross-cultural behavior. Our existing data do not allow us to fully address these issues.

Comprehensiveness of the 14 Categories of Behavior

The survey concluded by asking for written comments on additional skills not represented in the survey. Of the approximately 6,300 people who viewed this query, 523 commented.

The comments fit into three general themes:

- nomination of an additional skill or skills (a valid response that addressed the intent of the question)
- agreement or disagreement with one or more of the 14 skills (a valid response that did not address the intent of the question)
- general comments about aggravation with deployment issues and perceived inequities, including some that were not flattering to the military, the Air Force, foreign cultures, or specific personnel. Comments of this type were few in number.

All 523 responses were first examined to develop a general sense of the common comments as a starting point for coding. Multiple discussions were held to revise the coding scheme to accommodate additional common comments. (For more detailed descriptions of the variables for which we coded, see Appendix G.) Two raters independently coded all comments for any content that applied to the coding scheme and then compared their coding results, discussed areas where they differed, and resolved any discrepancies. The results, presented in Table 3.8, are arranged in descending order of frequency of occurrence.

Table 3.8
Themes Found in Narrative Comments on Survey

Comment Theme	Number of Comments	Percentage of All Comments
Understand the culture (in general)	117	22
Not applicable—I worked inside the wire	62	12
Understand religious differences	41	8
Learn the culture of the TCNs	28	5
People skills/interpersonal skills	26	5
Combat skills (including weapons, convoys, survival skills, etc.)	23	4
Joint operations	22	4
Role of interpreters/using interpreters	20	4
Allied operations	19	4
Better training/general training	14	3
Business practices	13	2
Understand U.S. policy/mission	13	2
Force protection—awareness (identify/avoid suspicious activity)	12	2
Educate/train foreign military	7	1
How to drive in the country (not combat related)	6	1

NOTE: Any response that addressed more than one coding category could be recorded in multiple categories.

By far the most common type of comment spoke of a general need to understand the culture of the deployment location. Many of those who commented also indicated that they worked inside the wire and had limited contact with host nationals. Despite the many comments, the coding did not reveal any cross-cultural performance categories not already covered by the existing 14. Some types of comments—such as "understand the culture"—overlapped with existing categories; other types—such as combat skills—did not fall under the heading of cross-cultural performance at all; instead, they echoed the need for predeployment combat training described in a recent RAND report by Manacapilli, Hardison, Gifford, Bailey, and Bower (2007).

We did, however, find two categories worth noting. Several people commented on the role of TCNs and allied operations as areas applicable to cross-cultural training. More specifically, several people indicated that during deployment they regularly interacted with TCNs (e.g., individuals from countries neighboring the deployment country who worked and lived there) and would benefit from an understanding of the TCN's culture in addition to or instead of the deployment country's culture. Others indicated that they regularly worked with allied forces whose cultures were foreign to them and would benefit from training in those cultures. While the 14 categories address training in the cultures of allied forces and TCNs, it is important to recognize that these cultures are not always the culture of the country of deployment. In such cases, cross-cultural training would have to be tailored to the cultures of the TCN and allied forces in question in order to address the comments raised by survey participants.

Some comments included specific agreement or disagreement with the 14 behavior categories proposed in the survey. Table 3.9 shows the numbers of positive (i.e., agreeing with the usefulness or importance of a behavior category) and negative (i.e., disagreeing with the usefulness or importance of a behavior category) comments that were made. As can be seen, comments on the importance of respecting cultural differences were by far the most common,

Table 3.9
Positive and Negative Survey Comments About the 14 Behavior Categories

Category of Behavior	Positive Comments		Negative Comments	
	Number	Percentage	Number	Percentage
Gathering and Interpreting Observed Information	24	5	3	1
Applying Appropriate Social Etiquette	13	2	6	1
Respecting Cultural Differences	92	18	5	1
Self-initiated Learning	14	3	4	1
Changing Behavior to Fit Cultural Context	13	2	4	1
Establishing Credibility, Trust, and Respect	22	4	3	1
Verbal and Nonverbal Communication	20	4	3	1
Managing Stress in an Unfamiliar Cultural Setting	18	3	4	1
Negotiating with Others	14	3	5	1
Applying Regional Knowledge	50	10	3	1
Resolving Conflict	7	1	5	1
Foreign Language Skills	77	15	12	2
Influencing Others	9	2	5	1
Establishing Authority	7	1	10	2

echoing the high frequency of comments on the theme of general cross-cultural understanding (see Table 3.8). Categories with a relatively higher number of negative comments showed a tendency to be those with lower importance ratings, such as foreign language skills and establishing authority.

Three examples of comments calling for more or better training that highlight the need for more training in general were as follows:

- "Integrate training into daily ops. Can't take a 2-hour on-line course and get it. Need to practice daily."
- "I had no training for job or country I was sent to."
- "Understand religious aspects in country, and how religion applies to everyday life of the population."

Some comments relating to foreign language skills described two issues (mentioned in Chapter One) with existing language training:

- "If USAF was serious about language skills it would make language programs easily available."
- "Send me to a country that actually speaks the language which I graduated the Defense Language Institute for. I graduated DLI in the [specific foreign language cited] . . . and have yet to use the language in the military."

Finally, some comments showed the importance with which some personnel view the need to respect cultural differences, whereas a few other comments showed a strong lack of respect for cultural differences. Both of these views, the positive and the negative, support the need for training in this category. The following comments illustrate this need:

- "In over 7 months of teaching Iraqis and 28 days of doing the same in Afghanistan the most important deficiency that must be overcome is a sense of superiority on the part of the USAF and USA. Iraqis and Afghanis aren't stupid, just different."
- "I don't want to learn to speak, write or communicate with foreigners. That's what we have interpreters for. Don't create more work for us. If we are in another country helping their situation, they need to learn to adapt to our language, culture, and the way we live [and] be happy with it. The less contact I have with foreigners, the better."

CHAPTER FOUR

Conclusions and Recommendations

The first of our four goals for the survey was to verify the importance of the original 14 categories of cross-cultural behaviors. The results of the survey show that each of those 14 behaviors is viewed as important by at least some airmen (see Figure 3.1).[1] The results also show that certain categories are viewed as more important, on average, than others. This suggests that all categories should be included in training for all airmen, and that categories of higher importance should be emphasized more than those of lower importance.

Our second goal was to identify whether there were categories of cross-cultural behavior other than the original 14. To perform this identification, we coded the open-ended responses on the survey. We found that respondents had suggested several areas for additional training, but that these either fell within our existing 14 categories of behavior (e.g., interacting with TCNs and interacting with coalition forces) or were not cross-cultural behavior at all (e.g., weapons skills and convoy driving). As a result, we identified no new category of cross-cultural behavior to add to our existing 14.

Our third goal was to determine whether training needs differed by AFSC, grade (enlisted and officer), and deployment location. We found that level of importance ratings differed across AFSCs: Certain AFSCs (e.g., Security Forces, FAOs, and Contracting) agreed or strongly agreed that nearly all categories were important, some AFSCs (e.g., Intelligence, Aircrew Protection, and Aircraft Maintenance) somewhat agreed that most categories were important, and other AFSCs tended to neither agree nor disagree or slightly disagreed that any categories were important (e.g., pilots and navigators). In addition, unique patterns in the importance of certain categories were observed within certain AFSCs. For example, Contracting showed noticeably stronger agreement with the importance of negotiating with others than it did with the other categories.

---

[1] No items were included to assess the possibility of false positives from accidental (i.e., random) endorsement of a given category. However, the results of our AFSC-specific analyses allow us to rule out the possibility that random error was solely responsible for all categories being viewed as important by some airmen. If one of the categories at the bottom of the importance ranking had been accidentally endorsed as important when it was important to no one, the accidental endorsement would most likely be randomly dispersed across AFSCs. More specifically, the within-AFSC mean scores for that category would be low for all AFSCs and not noticeably or significantly higher for any one AFSC relative to other AFSCs. In fact, quite the opposite occurred. Certain AFSCs had noticeably and significantly higher mean importance ratings than did other AFSCs for those categories viewed by the overall sample as least important. To illustrate, the overall sample viewed establishing authority as the least important of the categories. However, Security Forces rated this category as important for their deployed job, and their ratings for this category were noticeably higher than those of most other AFSCs. Therefore, establishing authority is viewed as not important by most airmen but is viewed as important by Security Forces. And this pattern held: All categories were endorsed more strongly by some AFSCs than by others. This finding, of differing levels of endorsement by AFSC, suggests that all 14 categories should be included in any comprehensive definition of cross-cultural behavior used by the Air Force.

Our final goal was to determine the perceived helpfulness of current Air Force training in the 14 categories. The results of the questions about training helpfulness suggest that many deployed airmen either are not receiving training or are unaware of receiving it in each of the categories. Respondents who did report receiving training were slightly more likely to report that they had been trained at some point in their career rather than just prior to deployment, and many of them did not view the training as helpful. This suggests that many airmen believe that existing Air Force training in all 14 categories is in need of improvement.

Three Levels of Training

The findings and observations discussed in this report highlight the importance of training airmen in cross-cultural behaviors for their deployed jobs. However, the finding that the importance assigned to the 14 categories of behavior differs significantly across AFSCs and, in some cases, by grade and deployment location leads us to conclude that the amount of cross-cultural training received should not be the same for all airmen. Since attempting to tailor training programs to each combination of AFSC, grade, and deployment location would be impractical, we suggest that any comprehensive cross-cultural training program provide at least three graduated levels of training.

The lowest level of training could include some minimum amount of cross-cultural training in each of the 14 categories for all airmen. A recent study found that many deployed airmen in a variety of non-combat AFSCs often and unexpectedly find themselves traveling or working outside the secure base (Manacapilli, Hardison, Gifford, Bailey, and Bower, 2007). We therefore suggest that this lowest level of cross-cultural training be provided to all airmen as part of their PME and during predeployment training. This will ensure that all airmen have a basic foundation of cross-cultural behavior ready to apply if the need arises. This minimum-level training could focus more heavily on the behavior categories that, on average, were rated most important by the survey respondents (see Figure 3.1 for the categories' ranked order of importance). However, no category should be neglected.

The next, medium level of training could provide cross-cultural behavior training of greater depth to airmen in the AFSCs that indicated that the 14 categories were, on average, moderately important. As with the low-level training, no category would be left out, but attention would be paid to categories generally rated as more important than the others.

The final, highest level of training would be provided to airmen in those AFSCs that indicated clear and strong agreement with the importance of a particular category. For these select AFSCs, this specialized, expert-level training would be only in categories whose importance they had strongly agreed with. Thus, in contrast to the first and second levels of training, this level would be category specific by AFSC.

Appendix E's Tables E.1 and E.2 show the importance ratings by AFSC, grade, and deployment. These two tables could be used to identify the AFSCs that are in need of more than the lowest level of training—i.e., those needing the moderate (medium-level) training, and those also needing the intensive (high-level) training in specific categories.

Next Steps

To confront the ongoing challenge of building a cross-culturally skilled force, we recommend further research and development on training for cross-cultural performance.[2] We view the process of building a cross-culturally skilled force as an instructional system and as needing to be approached comprehensively.

The suggestions we make here represent interdependent steps that will result in a coherent, institutionalized instructional system that takes into account the organizational context of the Air Force. These suggestions also represent best practices in training (e.g., Goldstein, 1991). This type of systematic approach would facilitate lifelong learning of behaviors identified as useful to cross-cultural performance; it would also ensure that the content of cross-cultural training would be successfully transferred from the training context to the work context and maintained. We have organized our suggestions in roughly temporal order, and we identify areas in which further research would be helpful in determining the best course of action.

Set Performance Standards

Our study did not address the issue of absolute performance standards. This issue, which is more than simply measuring or tracking training, has to do with establishing the actual amount of skill that must be displayed on a job in order to succeed. While the higher levels of importance ratings given to categories by some AFSCs do suggest that those AFSCs need higher levels of skills, no minimum performance standard was specified. So, for example, even though some AFSCs indicated that foreign language skills in the country of deployment are important, there is no way to know what skill level (e.g., novice) would be sufficient. And even if a level were indicated, the meaning of the level (e.g., expert) for each behavior would not be clear. In other words, although our findings may offer some guidance, additional research is needed to establish standards for performance for each of the 14 categories.

DoD has begun to chart a course in this direction with its development of guidelines for regional and cultural expertise (DoD, 2007). However, our examination of these guidelines, available in draft form, revealed that they largely describe in-depth expertise in only one type of cross-cultural behavior—applying regional knowledge. More specifically, DoD has provided descriptions of levels of regional knowledge, from pre-novice through expert. The person has only very limited knowledge of basic facts about the country at the lowest levels of expertise, and the person has extensive knowledge about the country (e.g., the political structure of the country) at the highest levels.

Expansion of the DoD guidelines to cover all 14 behavior categories could bring more breadth and completeness to the determination of expertise, since the five skill levels outlined in the guidelines could be delineated separately for each of our 14 categories of behavior. The DoD guidelines could also provide more depth to the conceptualization of our 14 categories by considering behaviors that fall at a lower level of expertise than those in our current descriptions of the categories. The application of DoD's skill-level framework to each of the 14 categories is a suggested next step for establishing culture performance standards.

As an example of how the DoD guidelines could be combined with our 14 categories of behavior, consider self-initiated learning. Level 0+ (pre-novice) would describe a person who might know information is available, but does not know where or how to seek that informa-

2 As we did throughout this report, we use *training* to refer to both *training* and *education* in the discussion.

tion. Level 1 (novice) would describe a person who is able to refer to briefing notes and possibly class notes for information on the culture, and knows where to seek further knowledge but has insufficient information to evaluate the quality of the content of his/her sources. Level 2 (associate) would describe a person who has examined some sources independent of those provided in area studies courses and other military education, and who has sufficient information to make judgments about the quality of the information. Level 3 (professional), in contrast, would describe a person who has investigated the culture on his/her own, bought a few books, and sought out information on the Internet. This person could very easily apply this ability to search for knowledge to his/her job. Level 4 (senior professional) would describe someone who has gone out of his/her way to take courses in the culture and seek out experiences with members of the culture and exposure to host nation media outlets in order to gain greater knowledge. Level 5 (expert) would describe someone who has spent the majority of his/her free time with host nation individuals and has sought full-immersion experience in order to understand the culture in the greatest depth.

Develop Training Content to Address Each of the 14 Behaviors

An in-depth examination of current Air Force cross-cultural curricula and a comparison of these curricula with other services' programs (including select business models) could be used to help Air Force trainers build on existing effective programs and to ensure content coverage for each of the 14 categories of behavior identified as important. This undertaking would help identify current gaps in available Air Force training. If training programs for each of the 14 categories are already available in the other military services or elsewhere, they could be repurposed for the Air Force. An examination of extant Air Force training programs (which, as we learned from our informal interviews and focus groups, are at present offered unsystematically throughout the force) is also essential. These existing Air Force programs may be especially useful in that they would not have to be modified to the same extent as programs developed for students in other services or for civilian audiences. To the extent that existing training programs do not address the 14 categories, new content will have to be developed.

Another issue for development of training content is the required level of cultural specificity. Some of the 14 categories of behavior represent somewhat culture-general behaviors. For example, the category of self-initiated learning and the category of gathering and interpreting observed information are more likely to cover behaviors that are applicable regardless of host culture. However, other categories, such as applying regional knowledge and negotiating with others, deal with content that is more culture specific. Though this study identified 14 categories of behavior needed by Air Force personnel and identified the content of all categories in general terms, it did not address the content of training that would be needed to prepare airmen to effectively perform some of those behaviors in a specific region or country.

To illustrate, while negotiation is a skill needed by contracting personnel regardless of deployment location, negotiating techniques differ across cultures.[3] Airmen who are expected to negotiate in, say, Afghanistan need to receive training on the techniques for successful negotiation in Afghanistan, including any of Afghanistan's principle subcultures. Ideally, airmen would learn effective and ineffective negotiation tactics, the different stages of negotiation, etc. Our study did not specify appropriate negotiation tactics, stages of negotiation, or any other

[3] See Ting-Toomey, 2004, for one approach to possible differences, and see Gelfand and Dyer, 2000, for a model of cross-cultural negotiation.

relevant information needed for effective negotiation in any specific culture or cultures.[4] We therefore suggest that further research address the details of country- or region-specific training in each of the 14 categories. Useful sources for this type of information would include cultural anthropologists with extensive experience with the region and natives of the region.

As part of an Air Force–wide instructional system, this culture-specific training would likely be delivered in the context of predeployment training rather than PME. PME content, in contrast, might be expected to focus on general negotiation tactics, how to determine when culture-specific negotiation tactics might apply, and how to provide a broad overview of different culturally appropriate negotiation tactics. The goal would be to instill awareness of the issues and foundational knowledge that can be built on through specific predeployment training. Further research and consultation with current best practices would determine the specific content of the PME curricula.

For example, research suggests (Glaser, 1984) that learning is a cumulative process in which individuals try to integrate new knowledge with what they already know. Acquisition of new material is thus facilitated when it is part of a cumulative process in which the new material is integrated into an existing foundation of knowledge or frame of reference. This research suggests that predeployment training itself would be more effective if it were built on such an existing foundation, perhaps acquired through PME. Training delivered in the context of PME could build on a foundation started in Basic Military Training. Research also suggests that simple accumulation of expert knowledge is not sufficient to ensure improved performance. Instead, understanding must be acquired through a time-consuming process of activity and experiences, starting with basic concepts that can form the basis for expert understanding and application (Greeno, Collins, and Resnik, 1996). Sequential levels of training over one's career would therefore be helpful in allowing this growth of expertise in each of the 14 categories of behavior.

Similar concepts could be applied to specialized training for certain career fields. However, for career fields that specialize in a specific skill (such as negotiation in the career field of Contracting), it would be advisable for PME to also include culture-specific training so that airmen in that field do not rely solely on predeployment training for culture-specific tactics associated with their specific skill.

We have no data on which to base suggestions for the appropriate order in which to provide training in the 14 categories. Nevertheless, the acquisition of skills in some categories seems to naturally precede the acquisition of skills in others. For example, communication is logically a necessary component of establishing influence, establishing trust, and effective negotiation. It is therefore sensible that some training in culture-specific communication should precede culture-specific goal-oriented behaviors. Other natural orders of presentation are less clear. For example, it is not clear that foreign language skills need to precede training in any of the other 14 categories. Ultimately, the extent to which skill in an area depends on skill in another should be the determining factor for sequencing the training of the 14 categories. We thus suggest that the appropriate order for training the categories be further explored.

We also have no data on ease of training for the 14 categories. For example, it is possible that simply forewarning people about the types of situations they may encounter that typically

4 Note that one country or region does not represent only one culture. Within any country or region, multiple subcultures may be relevant for training purposes. Ultimately, training needs to reflect differences in appropriate behavior among subcultures.

induce stress could help mitigate the stress of such situations. However, it is also possible that no meaningful change in situation-associated stress levels can be brought about without intensive and repeated exposure to such situations. Also, it is known that learning a foreign language requires extensive time and resources, but the amount of language skill that would translate to a large impact on cross-cultural performance is not clear. Further research on the impact of equivalent training resources on performance in each of the 14 categories is needed.

Finally, basic principles of the learning and transfer of material should be applied in designing the training curricula. For example, the literature examining the factors that lead to successful transfer of skills acquired through training to on-the-job performance suggests several findings that should be considered. For example, training's transferability may depend on the extent to which training is high fidelity (i.e., resembles the situations actually encountered on the job and reflects the variability of situations encountered on the job), its content concentrates on general principles (i.e., is not situation specific), and it provides multiple opportunities to practice (Kraiger, 2003).

Tools for Measuring Skills of Air Force Personnel

Tools for measuring skill levels in each of the 14 categories of behavior could be developed and ultimately used to evaluate both the skills of airmen and the success of training. The type of measurement tool should depend on both the type of behavior to be assessed and the level of skills to be detected. For the lower skill levels, cost and time considerations favor the use of multiple-choice tests. For example, lower levels of applying regional knowledge could be measured using a paper-and-pencil test of factual information. Higher levels could be measured by a paper-and-pencil test that requires test takers to apply that knowledge in making strategic decisions. Similarly, situational judgment tests (also multiple-choice tests) could be used to quickly and inexpensively evaluate lower-level skills for categories such as changing behavior to fit cultural context and applying appropriate social etiquette. These tools would likely be most applicable for those AFSCs that ranked lower in the importance ratings, on average, for each of the categories.

More-expensive and higher-fidelity modes of skill measurement would be most appropriate for evaluating the highest levels of skills in several of the categories and would be applicable for the skill sets in those AFSCs that had higher importance ratings, on average, for the 14 categories. Tests similar to assessment centers (multiple scenario-based work samples, often involving role-play exercises) could be used to evaluate expert-level performance in most of the categories. This is functionally equivalent to the type of oral testing currently used to evaluate foreign language skills.

As noted above, training content should be region and country specific; testing should reflect this. For example, rules for negotiation in Japan are very different from those in Afghanistan. Tests should be designed to assess only those skills appropriate to a given culture.

Evaluate the Success of Current Training Efforts Against the Standards

Existing training (as well as the to-be-developed training suggested by the 14 behavior categories we examined) should be evaluated to determine how much improvement it produces in each of the 14 categories. One method of evaluating training is to use reaction criteria. Specifically, this involves asking trainees for their opinions of the training at the end of the training program (Alliger and Janak, 1989). This is often used as one—and sometimes the only—method of evaluating cross-cultural training programs (Mendenhall, Stahl, Ehnert, Oddou,

Osland, and Kulhmann, 2004). Although this form of evaluation in isolation is clearly not sufficient, since the relationship between reactions and on-the-job behavior is not large, research indicates that asking trainees to indicate the perceived utility of training may be more useful than measures of learning during training for evaluating the potential to impact job behavior (Alliger, Tannenbaum, Bennett, Traver, and Shotland, 1997). Hence, this approach may be useful in the initial stages of evaluating the training programs.

Our study results on respondents' perceptions of the usefulness of training, which are examples of reaction criteria, suggest that the current Air Force training would benefit from improvement. It should be noted that several caveats apply to this finding, however, including that the participants decided what they considered *training* and that reactions were not collected at the conclusion of a particular experience and thus cannot be considered applicable to a specific program. However, the results do confirm the global assessment of Air Force leadership that training for cross-cultural performance represents an area for improvement.

The tools developed to measure skill levels (discussed above) could be used as another evaluation method and would address what is discussed in the literature as "learning" criteria (i.e., a measure of knowledge acquired in training or education). Some of these suggested measurement tools are quite inexpensive to employ once developed. Given ease of use, such tools offer the potential to do more than simply measure knowledge acquisition at the end of a training program. They may also be used over time to determine whether the required knowledge is maintained. Knowledge of how long skill levels are maintained would be useful for other tasks, as well, such as scheduling refresher courses.

Another strongly encouraged method of evaluating the success of training programs is to assess trainees' performance in each of the 14 categories of behavior before and after training. In this way, one can examine improvement as a result of training and can determine whether minimal performance standards have been met. In the academic literature, this is discussed in terms of whether training transfers to the job performance context. Although the academic literature focuses primarily on supervisory or peer performance ratings when performance criteria other than self-ratings of performance are examined (e.g., Mendenhall, Stahl, Ehnert, Oddou, Osland, and Kulhmann, 2004), ideally, performance ratings from host nationals would be considered.

Although the cross-cultural training literature often uses reaction criteria or learning criteria (i.e., cultural knowledge) to evaluate the success of training programs (Mendenhall, Stahl, Ehnert, Oddou, Osland, and Kulhmann, 2004) these types of criteria do not always translate to successful job performance (Alliger, Tannenbaum, Bennett, Traver, and Shotland, 1997; Goldstein, 1991). Similarly, as discussed in Chapter Three, our results are based on airmen's perceptions of the helpfulness of training and thus should be viewed with caution. Again, we think it is essential that other types of criteria for evaluating current Air Force cross-cultural training be examined.

Finally, other measures of success targeted to categories of behavior might perhaps be considered. For example, for the category of negotiating with others, the costs of contracts negotiated by airmen trained versus untrained in culturally appropriate negotiation skills might serve as a relatively concrete and easily obtainable criterion for success. Such a criterion would fall into what is known in the literature as results criteria (i.e., indications of how training impacts the organizational bottom line).

Track Airmen's Skills and Training

Once there are adequate tools for measuring skills and the training programs for the 14 behaviors have been identified, the tracking of airmen's skills and training received will be a critical step in building a cross-culturally proficient force. Existing systems for tracking the training of airmen should be examined to determine whether they capture current training in cross-cultural behaviors.

Tracking the training that airmen receive is essential to any thorough evaluation of training, because record keeping permits evaluation on the basis of job performance, promotions, and even mission success years later. Without a record of who has participated in which training programs, long-term evaluation of the Air Forces training and development efforts will be impossible.

Tracking the skill levels of airmen is critical for the selection and placement of personnel. To the extent that cross-cultural skills are measured and recorded, they can permit more systematic and informed deployment assignments, identification of airmen with pre-existing aptitudes for cross-cultural performance, and selection of qualified airmen to receive specialized cross-cultural training. Lastly, the tracking of current skill sets can prevent placement of airmen into training programs that would be redundant or below their skill level.

Thus, attention to the tracking process would enable the Air Force to allocate its resources most efficiently. In addition, it would enable the Air Force to clearly identify the airmen who are most able to benefit from cross-cultural performance training and to use it most effectively during performance of their deployed jobs.

Establish the Appropriate Availability of Training

We suggest that Air Force–wide training over the course of airmen's careers focus on general principles, and we recommend that predeployment training be the venue for information that is more situation specific. In this way, airmen will have multiple opportunities to acquire more general information and practice throughout their careers. We also suggest that there be multiple levels of cross-cultural performance training, as described earlier, and that even airmen who are highly unlikely to be exposed to cross-cultural interaction receive a general grounding (the lowest level of training) in all 14 cross-cultural behavior categories.

Training can be expensive, so it is crucial that the right training be available to those that need it and that training efforts provide a good return on investment. This is why evaluation of training's effectiveness is itself a wise investment. One way to ensure a high return on investment is to make the most-expensive and time-consuming training available only to those who need it. Examples of such training include classroom training to speak a language fluently, culture-immersion programs, and extensive, real-world culture simulation training. Another way to achieve a high return on investment is to make relatively inexpensive training programs available to as many people as possible. Examples of inexpensive training include computer-based training and training via books, online resources, videos, and other electronic media. These types of training resources could be designed specifically to address the 14 categories or compiled using any existing resources that address some of the 14 categories. These less-expensive training modes would perhaps be good vehicles for the lower levels of training.

Institutionalize Training (Systematic Effort)

A final key to incorporating a successful instructional system is for leadership to demonstrate that cross-cultural performance is an integral part of performance for all airmen. Suggested possibilities along these lines include ensuring that cross-cultural performance (as defined by the 14 categories) is more fully represented in the Air Force's Institutional Competency List. Another possibility is to provide, as recommended earlier, a foundation of cross-cultural training in Basic Military Training and officer commissioning training that is continuously reinforced and built on with training that occurs across each airman's career. The less-expensive modes of training (see previous subsection) might also prove useful for self-initiated practice programs. Yet another possibility is a system of organizational rewards or other incentives[5] that would convey the message that the Air Force values the practice and incorporation of cross-cultural training and thus encourage trainees to maintain and apply the skills they acquire in the 14 categories.

The learning of complex skills requires a comprehensive approach, and a needs assessment, such as we have provided here, is only the first step. Our findings represent a foundation on which to build an integrated approach to the Air Force's training of cross-cultural performance behaviors.

---

[5] See, for example, Kraiger's "Perspectives in Training and Development" (2003).

APPENDIX A

Survey Population and Sample

This appendix presents, in tabular form, information on the deployed population, the sample invited to participate in the survey, and the Air Force population sizes in all stratification subgroups.

Deployed Population, Sample Size, and Air Force Population

	Number of Personnel Deployed in Previous 18 Months						Size of Invited Sample						Size of Air Force Population		
	High Grade		Low Grade		Total		High Grade		Low Grade		Total		Grade		
2-Digit AFSC	Other	IRAQ/ AFG	Other	IRAQ/ AFG	Other	IRAQ/ AFG	Other	IRAQ/ AFG	Other	IRAQ/ AFG	Other	IRAQ/ AFG	High	Low	Total
10	32	18	0	0	32	18	32	18	0	0	32	18	301	0	301
11	962	392	2,711	700	3,673	1,092	150	150	150	150	300	300	5,324	6,586	11,910
12	372	161	900	254	1,272	415	150	150	150	150	300	300	1,639	1,861	3,500
13	184	70	413	153	597	223	150	70	150	150	300	220	1,696	3,038	4,734
14	108	68	427	236	536	306	108	68	150	150	259	220	1,031	1,686	2,717
15	9	15	38	37	47	52	9	15	38	37	47	52	243	358	601
16	43	39	5	6	50	45	43	39	5	6	50	45	1,085	60	1,145
20	10	18	0	0	10	18	10	18	0	0	10	18	168	0	168
21	153	133	470	295	626	428	150	133	150	150	303	283	1,364	2,006	3,370
30	26	26	0	0	26	26	26	26	0	0	26	26	301	0	301
31	12	45	50	178	62	223	12	45	50	150	62	195	272	405	677
32	48	98	139	218	188	316	48	98	139	150	188	248	508	698	1,206
33	94	91	397	317	504	410	94	91	150	150	257	243	1,233	1,997	3,230
34	18	15	87	45	105	60	18	15	87	45	105	60	130	349	479
35	25	11	40	32	65	43	25	11	40	32	65	43	116	169	285
36	9	15	41	26	52	41	9	15	41	26	52	41	0	856	856
37	31	36	83	51	115	87	31	36	83	51	115	87	580	856	1,436
38	3	1	13	3	16	4	3	1	13	3	16	4	0	0	0
40	14	7	0	0	14	7	14	7	0	0	14	7	104	0	104
41	37	44	44	40	81	84	37	44	44	40	81	84	559	495	1,054
42	17	34	48	70	65	104	17	34	48	70	65	104	475	549	1,024
43	42	49	34	32	76	81	42	49	34	32	76	81	621	535	1,156
44	77	120	48	46	125	166	77	120	48	46	125	166	1,086	896	1,982
45	38	113	4	16	42	129	38	113	4	16	42	129	459	263	722
46	94	99	269	258	363	357	94	99	150	150	244	249	1,226	273	1,499

2-Digit AFSC	Number of Personnel Deployed in Previous 18 Months						Size of Invited Sample						Size of Air Force Population		
	High Grade		Low Grade		Total		High Grade		Low Grade		Total		Grade		
	Other	IRAQ/ AFG	Other	IRAQ/ AFG	Other	IRAQ/ AFG	Other	IRAQ/ AFG	Other	IRAQ/ AFG	Other	IRAQ/ AFG	High	Low	Total
47	24	12	8	4	32	16	24	12	8	4	32	16	512	342	854
48	81	51	57	26	138	77	81	51	57	26	138	77	422	158	580
51	49	43	37	56	86	99	49	43	37	56	86	99	612	600	1,212
52	25	49	43	43	68	92	25	49	43	43	68	92	256	300	556
60		1		0		1		1		0		1	58	0	58
61	16	4	15	9	35	13	16	4	15	9	35	13	282	558	840
62	5	10	31	21	39	32	5	10	31	21	39	32	290	1,879	2,169
63	17	42	31	33	49	75	17	42	31	33	49	75	1,236	1,045	2,281
64	4	52	69	111	73	163	4	52	69	111	73	163	302	497	799
65	17	37	22	43	57	84	17	37	22	43	57	84	290	422	712
71	4	14	17	30	30	64	4	14	17	30	30	64	117	208	325
80	1		0		1		1		0		1		49	0	49
81	8	8	6	4	14	12	8	8	6	4	14	12	364	425	789
82	3	3	1	1	4	4	3	3	1	1	4	4	69	46	115
83	0	0	2	4	2	4	0	0	2	4	2	4	50	63	113
85	1		0		1		1		0		1		1	5	6
86	5	3	3	2	8	5	5	3	3	2	8	5	48	56	104
87	1		0		1		1		0		1		49	0	49
88		0		1		1		0		1		1	20	16	36
90	11	10	0	0	11	10	11	10	0	0	11	10	295	0	295
91	18	13	0	0	18	13	18	13	0	0	18	13	463	2	465
92	12	7	65	22	77	29	12	7	65	22	77	29	1,368	3,951	5,319
95	1		0		1		1		0		1		0	0	0
96	0		2		2		0		2		2		2	18	20
97	7	12	6	3	13	15	7	12	6	3	13	15	194	61	255

	Number of Personnel Deployed in Previous 18 Months						Size of Invited Sample						Size of Air Force Population		
	High Grade		Low Grade		Total		High Grade		Low Grade		Total		Grade		
2-Digit AFSC	Other	IRAQ/ AFG	Other	IRAQ/ AFG	Other	IRAQ/ AFG	Other	IRAQ/ AFG	Other	IRAQ/ AFG	Other	IRAQ/ AFG	High	Low	Total
1A	1,431	633	1,248	218	2,679	851	150	150	150	150	300	300	5,416	4,069	9,485
1C	676	983	554	859	1,231	1,842	150	150	150	150	301	300	6,138	4,070	10,208
1N	437	313	436	290	874	603	150	150	150	150	301	300	5,402	6,253	11,655
1S	41	39	1	0	42	40	41	39	1	0	42	40	341	27	368
1T	290	118	276	115	566	233	150	118	150	115	300	233	1,354	1,239	2,593
1W	127	217	54	68	182	286	127	150	54	68	182	219	1,395	780	1,395
2A	5,923	2,396	5,727	2,505	11,660	4,906	150	150	150	150	310	305	29,897	24,209	54,106
2E	689	596	654	705	1,344	1,302	150	150	150	150	301	301	6,100	4,465	10,565
2F	709	264	668	209	1,378	473	150	150	150	150	301	300	2,206	1,641	3,847
2G	119	87	37	22	157	109	119	87	37	22	157	109	520	256	776
2M	7	22	39	59	46	81	7	22	39	59	46	81	1,185	757	1,942
2P	61	5	42	5	103	10	61	5	42	5	103	10	467	274	741
2R	100	35	92	42	192	77	100	35	92	42	192	77	863	569	1,432
2S	1,067	643	650	374	1,718	1,017	150	150	150	150	301	300	4,763	2,913	7,676
2T	1,275	1,219	1,207	1,232	2,483	2,451	150	150	150	150	301	300	6,581	5,868	12,449
2W	905	714	980	768	1,885	1,483	150	150	150	150	300	301	8,092	6,656	14,748
3A	756	539	329	220	1,087	760	150	150	150	150	302	301	5,734	2,530	8,264
3C	1,026	627	617	405	1,645	1,040	150	150	150	150	302	308	7,593	3,025	10,618
3E	1,919	1,826	2,039	1,839	3,959	3,665	150	150	150	150	301	300	8,760	7,115	15,875
3H	7	5	0	0	11	5	7	5	0	0	11	5	3	—	3
3M	762	268	703	208	1,465	476	150	150	150	150	300	300	2,243	2,125	4,368
3N	77	78	20	33	97	112	77	78	20	33	97	112	1,744	679	2,423
3P	1,633	2,206	2,508	3,442	4,141	5,648	150	150	150	150	300	300	9,279	14,945	24,224
3S	467	345	121	92	589	437	150	150	121	92	272	242	6,085	1,638	7,723
3U	—	—	—	—	—	—	—	—	—	—	—	—	1	—	
3V	60	81	35	54	95	135	60	81	35	54	95	135	—	—	0

2-Digit AFSC	Number of Personnel Deployed in Previous 18 Months						Size of Invited Sample						Size of Air Force Population		
	High Grade		Low Grade		Total		High Grade		Low Grade		Total		Grade		
	Other	IRAQ/ AFG	Other	IRAQ/ AFG	Other	IRAQ/ AFG	Other	IRAQ/ AFG	Other	IRAQ/ AFG	Other	IRAQ/ AFG	High	Low	Total
4A	250	189	155	157	405	346	150	150	150	150	300	300	2,607	2,123	4,730
4B	39	25	25	12	64	37	19	14	16	8	35	22	544	306	850
4C	23	40	30	21	53	61	12	22	22	15	34	37	384	317	701
4D	1	13	9	22	10	35	1	8	5	13	6	21	176	200	376
4E	47	21	23	14	70	35	17	10	17	8	34	18	644	413	1,057
4H	27	27	18	39	45	66	10	8	13	29	23	37	137	197	334
4J	4	24	0	4	4	28	3	10	0	4	3	14	214	124	338
4M	25	11	20	5	45	16	13	2	16	2	29	4	225	127	352
4N	532	468	274	271	806	739	150	150	150	150	300	300	3,989	2,752	6,741
4P	36	32	9	20	45	52	23	14	6	14	29	28	566	316	882
4R	30	28	14	30	44	58	11	17	9	21	20	38	533	348	881
4T	32	53	27	26	59	79	14	23	17	14	31	37	721	521	1,242
4V	8	5	8	2	16	7	3	3	3	2	6	5	188	91	279
4Y	37	36	38	30	75	66	24	19	26	20	50	39	1,367	1,060	2,427
5J	44	67	6	1	50	68	44	67	6	1	50	68	828	156	984
5R	33	48	21	24	54	72	33	48	21	24	54	72	325	129	454
6C	168	211	59	51	227	262	150	150	59	51	209	201	842	330	1,172
6F	156	118	103	55	260	173	150	118	103	55	254	173	1,491	919	2,410
6N	0		1		1		0		1		1		—	—	0
7S	81	121	0	0	82	121	81	121	0	0	82	121	949	7	956
8A	9	1	0	0	9	1	9	1	0	0	9	1	175	1	176
8B	15	16	0	1	15	17	15	16	0	1	15	17	855	6	861
8C	6		0		6		6		0		6		132	—	132
8D		11		0		11		11		0		11	61	—	61

	Number of Personnel Deployed in Previous 18 Months						Size of Invited Sample						Size of Air Force Population		
	High Grade		Low Grade		Total		High Grade		Low Grade		Total		Grade		
2-Digit AFSC	Other	IRAQ/ AFG	Other	IRAQ/ AFG	Other	IRAQ/ AFG	Other	IRAQ/ AFG	Other	IRAQ/ AFG	Other	IRAQ/ AFG	High	Low	Total
8F	189	141	0	0	189	141	150	141	0	0	150	141	1,125	—	1,125
8G		1		0		1		1		0		1	101	185	286
8M	100	52	89	22	189	74	100	52	89	22	189	74	320	262	582
8R	15	6	4	3	19	9	15	6	4	3	19	9	2,095	101	2,196
9D	10	9	1	0	11	9	10	9	1	0	11	9	230	9	239
9E	4	1	0	0	4	1	4	1	0	0	4	1	146	—	146
9G	23	25	0	0	23	25	23	25	0	0	23	25	355	—	355
9L	3		0		3		3		0		3		64	5	69
9S	18		5		23		18		5		23		264	195	459
9T	0	0	8	13	8	14	0	0	8	13	8	14	127	6,607	6,734
9U	—	—	—	—	—	—	—	—	—	—	—	—	2	1	
	25,304	18,152	26,691	18,023	52,080	36,224	5,964	5,927	5,608	5,387	11,657	11,363	173,430	148,589	322,019

NOTES: (1) High grade is E-5 through E-9 for enlisted personnel and O-4 through O-9 for officers; low grade is E-1 through E-4 for enlisted personnel and O-1 through O-3 for officers. (2) In some cases, totals include individuals for whom no grade information was available and therefore are slightly larger than the sum of the individual subgroup sizes.

APPENDIX B

Survey Instrument

Survey Control Number: SCN07-048 (valid until July 2008)

The RAND Corporation, a non-profit research institution in Santa Monica, California, is conducting this survey as part of a foreign culture and language research initiative conducted on behalf of the U.S. Air Force. The goal of this research is to develop a framework for training Airmen to be aware of and sensitive to cultures that they may encounter when deployed.

We selected you for possible participation in this study because of your past experience as a deployed member of the U.S. Air Force. Even if you have very little deployed experience, your responses are useful to us. Your participation consists only of completing this on-line survey.

Your participation in the study is completely voluntary. You may refuse to participate, or you may stop at any time and for any reason, without any penalty.

Your responses to this survey are confidential. We will use the information you give us for research purposes only. We will protect the confidentiality of this information, and will not disclose your identity or information that identifies you to anyone outside of the research project, except as required by law. We will not identify you in any reports we write. We will destroy all information that may identify you at the end of the study.

DEMOGRAPHICS

What is your paygrade?
(Paygrade List)

What is your gender?
Male
Female
Decline to answer

What is your current duty AFSC?

If your AFSC is not listed, select "other" from the bottom of the drop-down list and type your AFSC in the box below.

(AFSC List)

If your AFSC is NOT listed in the above drop-down list, please enter it here.

Insert TEXT

Have you ever been deployed outside the United States?
* (The asterisk means we require an answer to this question in order for you to proceed with the survey.)
Yes
No

If they indicated they have not been deployed they saw the following message:

> While we appreciate your interest, our questions are only about deployed experiences outside the United States. Because you have not been deployed outside the United States, you have reached the end of the survey.
>
> Again, we appreciate your willingness to participate and thank you for your time.

If they indicated they have been deployed they continued to the next question.

During your career, how many times have you been deployed outside the United States?
1–5 times
6–7 times
8–10 times
11–15 times
16 times or more

How many days were you deployed outside the United States during the last 20 months?

Less than 30 days
30 to 60 days
61 to 90 days
91 to 120 days
Over 120 days

MOST RECENT DEPLOYMENT

In which country were you last deployed outside the United States?

* (The asterisk means we require an answer to this question in order for you to proceed with the survey.)

Afghanistan
Antarctica
Bahrain
Colombia
Cuba
Cyprus
Djibouti
Ecuador
El Salvador
Germany
Greece
Guam
Guatemala
Honduras
Iraq
Japan
Kuwait
Kyrgyzstan
Netherlands Antilles
New Zealand
Nicaragua
Pakistan
Puerto Rico
Qatar
Saudi Arabia
Serbia and Montenegro
South Korea
Spain
Turkey
The UN

United Arab Emirates
Other, please specify

Insert TEXT

How long was your last deployment to ______?
Less than 30 days
30 to 60 days
61 to 90 days
91 to 120 days
Over 120 days

Have you been deployed to ______ more than once?
Yes
No

During your last deployment to ______, what was your deployed unit's mission?
Flying ops
Other ops
Medical
Base support
CAOC or AOC
Staff above Wing level

In your last deployment to ______, did you frequently operate in or plan for operations in another country or AOR?
No
Yes. If so, what country or AOR?

Insert TEXT

What was your paygrade when you were deployed in ______?

In what AFSC did you deploy to ______?

If your AFSC is not listed, select "other" from the bottom of the drop-down list and type your AFSC in the box below.

Insert TEXT

LANGUAGE SKILLS

My level of skill in the foreign language that I know best allows me to:

— Do nothing—I do not speak, write, or understand any foreign language
— Satisfy routine travel needs and minimum courtesy requirements
— Satisfy routine social demands and limited work requirements
— Speak the language with sufficient structural accuracy and vocabulary to participate effectively in most formal and informal conversations on practical, social, and professional topics
— Use the language fluently and accurately on all levels normally pertinent to professional needs
— Speak with the proficiency equivalent to that of an educated native speaker

My level of skill in the language of the country where I was last deployed (______) allows me to:

— Do nothing—I do not speak, write, or understand the language of the country where I was deployed
— Satisfy routine travel needs and minimum courtesy requirements
— Satisfy routine social demands and limited work requirements
— Speak the language with sufficient structural accuracy and vocabulary to participate effectively in most formal and informal conversations on practical, social, and professional topics
— Use the language fluently and accurately on all levels normally pertinent to professional needs
— Speak with the proficiency equivalent to that of an educated native speaker

IMPORTANCE OF CATEGORIES

The remaining questions are about your most recent deployment experience outside the United States.

You indicated that your most recent deployment experience was in ____________. Please answer the remaining questions with regard to only your most recent deployment experience.

Please indicate how strongly you agree or disagree with the following statements (or select "Don't know"):
[Participants indicated their agreement on a scale from strongly agree = 5 to strongly disagree = 1, or could select the option "I don't know"]

In _______, I found myself in situations (on and off duty) where it was or would have been important for me to:

— Take the initiative to volunteer for or request additional opportunities from the Air Force to learn about the culture and/or language of the country where I was deployed.
— Change my behavior depending on the gender of the local person(s) with whom I was interacting.
— Apply factual information about the country's current political situation.
— Change or deviate from SOPs and/or ROE to accommodate cultural issues.
— Observe the behavior of others around me to learn how the local people interact with each other.
— Gain the trust of the locals.
— Apply factual information about the current religious beliefs of the local people.
— Change local people's perception of Americans and the U.S. military.
— Provide leadership and guidance to the locals in order to achieve my mission.
— Use different tactics when negotiating with locals than I would when negotiating with Americans.
— Follow television and/or radio programming, such as news programs, in the language of the country where I was deployed.
— Learn on my own (i.e., through the Internet, books, software such as Rosetta Stone, or by talking to other people) about the country where I was to be deployed.
— Recognize and manage the stress of feeling that the local people may not welcome the presence of U.S. troops.
— Convince locals to respect my opinion.
— Demonstrate respect for the cultural differences between Americans and the locals.
— Prevent and/or resolve conflicts that may have arisen from an American disrespecting the cultural norms of locals.
— Avoid common American practices that may be offensive in certain cultural settings, such as speaking loudly, pointing, or using the left hand to pass an object.
— Demonstrate my authority to ensure compliance by the locals.
— Observe how the locals react to other locals to learn what produces positive reactions and what produces negative reactions.
— Recognize and manage the stress of feeling like a "fish out of water" due to cultural and ethnic differences.
— Take the initiative to talk to someone from the country where I was deployed (such as my interpreter or guide) to learn more about the local culture and/or language.
— Gain respect of the locals.
— Know and understand local customs, conventions, and norms for behavior.
— Recognize that local people often do not fit the stereotypes for their culture.
— Recognize and manage the stress of feeling that the local people might view me as a stereotypical American.
— Read signs, newspapers, flyers, or other materials printed in the language of the country where I was deployed.
— Observe how the locals react to other American military personnel to learn what produces positive reactions and what produces negative reactions.

— Adapt my behavior to match and/or complement the behavior of local people.
— Form mutually beneficial partnerships with local people.
— Behave in ways appropriate to local social etiquette (such as customs/manners used at a dinner party).
— Reach a compromise situation with locals that made both sides pleased with the result.
— Show locals that I would not betray their trust in me.
— Interact with locals in ways that might be awkward for an American but necessary to fit in with the culture.
— Speak any foreign language (different from the local language of the country where I was deployed).
— Recognize that everyone has stereotypes, but that I should try not to form opinions about the locals I met based on stereotypes.
— Apply different rules for etiquette when interacting with different types of locals (e.g., officials vs. tradesmen, men vs. women).
— Change local people's opinions.
— Apply culture-specific rules for negotiation.
— Communicate my message clearly to the locals.
— Apply factual information about the religious history of the local people.
— Manage and/or resolve conflicts between locals.
— Resolve conflicts between Americans and local people.
— Apply factual information about the country's economic history and current economic situation.
— Persuade local people to go along with or accept an idea.
— Apply factual information about the country's political history.
— Recognize that American culture is different from the local culture but not necessarily superior.
— Speak the language of the country where I was deployed.
— Exercise a leadership role with respect to locals.
— Use a translator to communicate with non-English-speaking people.
— Establish that I had authority over one or more local individuals.
— Adjust my behavior to intentionally convey information about my mood, demeanor, status, or intent to locals.
— Defuse a conflict between Americans and locals that may have arisen from either side not understanding the differences in cultural norms.
— Exercise judicial/law enforcement powers (such as resolving basic litigation disputes) over locals.
— Apply factual information about the foreign relations of a given country (e.g., whether a neighboring country is an ally or a foe).
— Follow the rules for etiquette that are unique to the culture in order to make the locals feel comfortable around me.
— Hire (with my own money) a local instructor, interpreter, and/or someone in the country where I was deployed to help me learn the language and culture.
— Spend my off-duty time with the local people in the country where I was deployed to learn their language, customs, values, and understand their behavior.

- Negotiate with the local people.
- Control or restrict the behavior of the local people.
- Treat each new local person I met as an individual who may be very different from the other locals I met.
- Respect the local peoples' values, opinions, and mindset.
- Clarify to a group of locals in no uncertain terms that I was in charge of the situation and that all of the locals would have to do as I instructed.
- Observe my surroundings to identify which local people appear to be acting inappropriately or suspiciously and which do not.
- Write in the language of the country where I was deployed.
- Understand subtle non-verbal communication (e.g., their body language) when interacting with local people.
- Recognize and manage the stress I felt in a situation that arose from my or another member of my unit's lack of language skills.
- Recognize and manage the stress I felt in situations that arose from my lack of understanding of the cultural norms in the country where I was deployed.
- Influence or persuade local people to behave or act in a certain way.
- Bargain with local people for supplies or resources.
- Gain credibility in the eyes of locals.

TRAINING

The next two pages ask about training you received from the Air Force. The first page asks about your training just prior to deployment. The second page asks about training over the course of your career in the Air Force.

Please indicate how strongly you agree or disagree with the following statements.
[Participants indicated their agreement on a scale from strongly agree = 5 to strongly disagree = 1, or could select "USAF didn't train me to do this"]

The training I received from the Air Force just prior to deployment in ______ helped me to:

1. Read, write, and speak the language of the country where I was deployed
2. Communicate effectively (both verbally and non-verbally) for the country where I was deployed
3. Apply appropriate social etiquette for the country where I was deployed
4. Manage stress in unfamiliar cultural settings in the country where I was deployed
5. Change behavior to fit cultural context of the country where I was deployed
6. Gather and interpret information gained through observation in country where I was deployed
7. Apply regional knowledge of the country where I was deployed
8. Learn on my own about the country where I was deployed

9. Avoid stereotypes and respect cultural differences between the country where I was deployed and the United States
10. Establish authority in the country where I was deployed
11. Influence others in the country where I was deployed
12. Negotiate effectively in the country where I was deployed
13. Establish credibility and trust with people in the country where I was deployed
14. Resolve conflict in the country where I was deployed

Please indicate how strongly you agree or disagree with the following statements (or select "USAF didn't train me to do this")

When I was deployed in ______, the training I received from the Air Force over the course of my career helped me to:

1. Read, write, and speak the language of the country where I was deployed
2. Communicate effectively (both verbally and non-verbally) for the country where I was deployed
3. Apply appropriate social etiquette for the country where I was deployed
4. Manage stress in unfamiliar cultural settings like the country where I was deployed
5. Change behavior to fit cultural context of the country where I was deployed
6. Gather and interpret information gained through observation in country where I was deployed
7. Apply regional knowledge of the country where I was deployed
8. Learn on my own about the country where I was deployed
9. Avoid stereotypes and respect cultural differences between the country where I was deployed and the United States
10. Establish authority in the country where I was deployed
11. Influence others in the country where I was deployed
12. Negotiate effectively in the country where I was deployed
13. Establish credibility and trust with people in the country where I was deployed
14. Resolve conflict in the country where I was deployed

CATEGORIES OF BEHAVIOR MISSING

The purpose of this survey is to establish the cultural skills needed by Air Force personnel when deployed to foreign countries. We have identified the following 14 categories of such skills. If there is a skill that you believe is important and not represented in our list, please describe and explain in the comment box below.

1. Read, write, and speak the language of the country where I was deployed
2. Communicate effectively (both verbally and non-verbally) for the country where I was deployed
3. Apply appropriate social etiquette for the country where I was deployed
4. Manage stress in unfamiliar cultural settings like the country where I was deployed
5. Change behavior to fit cultural context of the country where I was deployed
6. Gather and interpret information gained through observation in country where I was deployed
7. Apply regional knowledge of the country where I was deployed
8. Learn on my own about the country where I was deployed
9. Avoid stereotypes and respect cultural differences between the country where I was deployed and the United States
10. Establish authority in the country where I was deployed
11. Influence others in the country where I was deployed
12. Negotiate effectively in the country where I was deployed
13. Establish credibility and trust with people in the country where I was deployed
14. Resolve conflict in the country where I was deployed
15. Other skill (please describe):

Insert TEXT

END

Congratulations, you have reached the end of the survey!

Thank you for participating.

Survey Items Grouped by Scale

ENABLING BEHAVIORS

Foreign Language Skills

1. Speak the language of the country where I was deployed.
2. Speak any foreign language (different from the local language of the country where I was deployed).
3. Read signs, newspapers, flyers, or other materials printed in the language of the country where I was deployed.
4. Write in the language of the country where I was deployed.
5. Follow television and/or radio programming, such as news programs, in the language of the country where I was deployed.

Verbal and Nonverbal Communication

6. Communicate my message clearly to the locals.
7. Understand subtle non-verbal communication (e.g., their body language) when interacting with local people.
8. Adjust my behavior to intentionally convey information about my mood, demeanor, status, or intent to locals.

Applying Appropriate Social Etiquette

9. Behave in ways appropriate to local social etiquette (such as customs/manners used at a dinner party).
10. Change my behavior depending on the gender of the local person(s) with whom I was interacting.
11. Know and understand local customs, conventions, and norms for behavior.
12. Follow the rules for etiquette that are unique to the culture in order to make the locals feel comfortable around me.
13. ITEM EXCLUDED IN SHORT VERSION: Apply different rules for etiquette when interacting with different types of locals (e.g., officials vs. tradesmen, men vs. women).

Managing Stress in an Unfamiliar Cultural Settings

14. Recognize and manage the stress I felt in a situation that arose from my or another member of my unit's lack of language skills.
15. Recognize and manage the stress I felt in situations that arose from my lack of understanding of the cultural norms in the country where I was deployed.
16. Recognize and manage the stress of feeling like a 'fish out of water' due to cultural and ethnic differences.
17. Recognize and manage the stress of feeling that the local people might view me as a stereotypical American.
18. Recognize and manage the stress of feeling that the local people may not welcome the presence of U.S. troops.

Changing Behavior to Fit Cultural Context

19. Adapt my behavior to match and/or complement the behavior of local people.
20. Interact with locals in ways that might be awkward for an American but necessary to fit in with the culture.
21. Avoid common American practices that may be offensive in certain cultural settings, such as speaking loudly, pointing or using the left hand to pass an object.
22. ITEM EXCLUDED IN SHORT VERSION: Change or deviate from SOPs and/or ROE to accommodate cultural issues.

Gathering and Interpreting Observed Information

23. Observe the behavior of others around me to learn how the local people interact with each other.
24. Observe my surroundings to identify which local people appear to be acting inappropriately or suspiciously and which do not.
25. Observe how the locals react to other American military personnel to learn what produces positive reactions and what produces negative reactions.
26. Observe how the locals react to other locals to learn what produces positive reactions and what produces negative reactions.

Applying Regional Knowledge

27. Apply factual information about the country's economic history and current economic situation.
28. Apply factual information about the country's political history.
29. Apply factual information about the country's current political situation.
30. Apply factual information about the religious history of the local people.
31. Apply factual information about the current religious beliefs of the local people.
32. Apply factual information about the foreign relations of a given country (e.g., whether a neighboring country is an ally or a foe).

Self-Initiated Learning

33. Take the initiative to talk to someone from the country where I was deployed (such as my interpreter or guide) to learn more about the local culture and/or language.
34. Take the initiative to volunteer for or request additional opportunities from the Air Force to learn about the culture and/or language of the country where I was deployed.
35. Spend my off-duty time with the local people in the country where I was deployed to learn their language, customs, values, and understand their behavior.
36. Learn on my own (i.e., through the Internet, books, software such as Rosetta Stone, or by talking to other people) about the country where I was to be deployed.
37. ITEM EXCLUDED IN SHORT VERSION: Hire (with my own money) a local instructor, interpreter, and/or someone in the country where I was deployed to help me learn the language and culture.

Respecting Cultural Differences

38. Recognize that local people often do not fit the stereotypes for their culture.
39. Respect the local peoples' values, opinions, and mindset.
40. Treat each new local person I met as an individual who may be very different from the other locals I met.
41. Recognize that everyone has stereotypes, but that I should try not to form opinions about the locals I met based on stereotypes.
42. Recognize that American culture is different from the local culture, but not necessarily superior.
43. Demonstrate respect for the cultural differences between Americans and the locals.

GOAL-ORIENTED BEHAVIORS

Establishing Authority

44. Establish that I had authority over one or more local individuals.
45. Clarify to a group of locals in no uncertain terms that I was in charge of the situation and that all of the locals would have to do as I instructed.
46. Control or restrict the behavior of the local people.
47. Exercise judicial/law enforcement powers (such as resolving basic litigation disputes) over locals.
48. Demonstrate my authority to ensure compliance by the locals.

Influencing Others

49. Influence or persuade local people to behave or act in a certain way.
50. Persuade local people to go along with or accept an idea.
51. Exercise a leadership role with respect to locals.

52. Change local people's opinions.
53. Provide leadership and guidance to the locals in order to achieve my mission.

Negotiating with Others

54. Negotiate with the local people.
55. Bargain with local people for supplies or resources.
56. Apply culture-specific rules for negotiation.
57. Reach a compromise situation with locals that made both sides pleased with the result.
58. Form mutually beneficial partnerships with local people.
59. Use different tactics when negotiating with locals than I would when negotiating with Americans.

Establishing Credibility, Trust, and Respect

60. Gain the trust of the locals.
61. Gain respect of the locals.
62. Gain credibility in the eyes of locals.
63. Change local people's perception of Americans and the U.S. military.
64. Convince locals to respect my opinion.
65. Show locals that I would not betray their trust in me.

Resolving Conflict

66. Manage and/or resolve conflicts between locals.
67. Resolve conflicts between Americans and local people.
68. Prevent and/or resolve conflicts that may have arisen from an American disrespecting the cultural norms of locals.
69. Defuse a conflict between Americans and locals that may have arisen from either side not understanding the differences in cultural norms.

Item not used in a scale:
Use a translator to communicate with non-English speaking people.

APPENDIX D

Summary Statistics

This appendix provides the weighted summary statistics for the importance ratings of the 14 categories of behavior (including the original scales and shortened scales) and for the self-ratings of skill on the foreign language known best and the language of the deployment location.

Weighted Summary Statistics for Importance Ratings of the 14 Categories of Behavior and Self-Ratings of Skill in Best-Known Foreign Language and Language of Deployment Location

Category or Item	Mean	Std. Dev.	1	2	3	4	5	6	7	8	9	10	11	12	13	14	15	16	17	18	19	20
1 Foreign Language Skills	2.77	.98	.86																			
2 Item: Speak deployment language	2.97	1.21	.87	NA																		
3 Verbal/Nonverbal Communication	3.13	.98	.70	.68	.85																	
4 Apply Social Etiquette	3.35	.94	.64	.61	.83	.90																
5 Apply Social Etiquette—SHORT	3.36	.97	.65	.63	.84	.98	.91															
6 Manage Stress	3.11	.94	.68	.62	.81	.82	.81	.90														
7 Change Behavior	3.05	.91	.69	.62	.79	.88	.86	.81	.83													
8 Change Behavior—SHORT	3.18	.97	.66	.61	.80	.89	.88	.82	.96	.85												
9 Gather/Interpret Observed Info	3.41	.97	.64	.60	.83	.88	.88	.82	.82	.83	.88											
10 Apply Regional Knowledge	2.99	.94	.73	.64	.76	.77	.75	.76	.78	.76	.75	.93										
11 Self-Initiated Learning	2.93	.88	.81	.73	.76	.78	.78	.78	.80	.79	.77	.77	.81									
12 Self-Initiated Learning—SHORT	3.11	.96	.78	.72	.78	.81	.81	.79	.80	.79	.80	.77	.98	.83								
13 Respect Cultural Differences	3.41	.93	.59	.58	.80	.89	.90	.80	.81	.83	.85	.71	.73	.77	.93							
14 Establish Authority	2.60	.95	.63	.57	.68	.54	.53	.62	.60	.56	.59	.66	.65	.59	.48	.91						
15 Influence Others	2.75	.92	.74	.69	.81	.71	.70	.73	.75	.72	.71	.78	.77	.74	.64	.85	.91					
16 Negotiate with Others	3.01	.94	.75	.68	.87	.85	.85	.80	.84	.84	.82	.80	.81	.81	.79	.69	.84	.92				
17 Establish Credibility	3.11	.95	.73	.68	.85	.85	.85	.82	.83	.83	.84	.79	.82	.83	.81	.67	.84	.89	.92			
18 Resolve Conflict	2.83	.97	.72	.64	.78	.69	.70	.75	.73	.72	.73	.78	.75	.72	.65	.79	.87	.81	.80	.90		
19 Skill in Best Foreign Language	.85	1.21	.15	.15	.11	.12	.12	.10	.12	.11	.13	.15	.15	.16	.11	.08	.12	.13	.14	.12	NA	
20 Skill in Deployment Language	.31	.83	.21	.20	.12	.08	.10	.06	.05	.06	.08	.09	.11	.13	.12	.06	.10	.13	.13	.12	.31	NA

NOTES: (1) Weighted means and standard deviations are in first two data columns, and coefficient alphas are located on the diagonal. All other values shown are weighted bivariate correlation coefficients. (2) NA = not applicable.

APPENDIX E

Survey Results

To augment the discussion in Chapter Three, this appendix presents individual, detailed results from our analysis of the survey responses for every AFSC, grade, and deployment location subgroup. There are 10 tables here, in three sets:

1. Tables E.1 and E.2 form the first set; they show the respondents' average ratings of the importance of the 14 categories of behavior. The results for the importance ratings on the three shortened scales are also shown, as are the results for participants' ratings of the extra item about skill in speaking the deployment-location language. Table E.1 gives the results for officers; Table E.2, for enlisted personnel.
2. The second set comprises Tables E.3 through E.6, which present the proportion of respondents reporting having received training in the 14 categories. Tables E.3 and E.4 cover, respectively, officer and enlisted results for training received over careers; Tables E.5 and E.6 do the same for training received just prior to deployment.
3. The tables in the third set, E.7 through E.10, are just like Tables E.3 through E6 except that they show the respondents' ratings of the helpfulness of the training they received.

All of the tables are laid out the same way. The left section consists of three columns: 2-digit AFSC, deployment location, and grade; the middle section has columns for each category of behavior (including both the original and shortened scales where applicable) and a column for the single item: speak the language; and the right section has three columns: number of survey respondents included in the analysis, number in the sample invited to participate in the survey, and total population of airmen who had deployed in the previous 18 months.

Within each table, right below the headings, is a row showing the weighted averages for all deployed personnel for comparison purposes. Below that row, results are presented for every combination of AFSC, deployment location, and grade. Except for the averages in the "all deployed personnel" row, all estimates in the tables are unweighted and representative of the ratings of those who deployed in that particular AFSC, grade, and location.

Some rows in the tables are blank or partially blank even when the sample size is listed as greater than zero. This is because we chose to not publish results whenever we had fewer than five individual responses contributing to the estimate. We made this decision for two reasons: Results based on only a few respondents are very unreliable, and, for small samples, the level of detail in these tables is precise enough to enable near-identification of respondents (to whom we guaranteed confidentiality of responses).

For the importance-rating and helpfulness-rating tables (Tables E.1 and E.2, and Tables E.7 through E.10, respectively) the results are color-coded to show the degree to which respondents agreed that the category was important/helpful (where 1 = strongly disagree and 5 = strongly agree). Light blue represents "mild" importance/helpfulness ratings (ranging from 3.1 to 3.4), medium blue represents "medium" importance/helpfulness ratings (ranging from 3.5 to 3.9), and dark blue represents "high" importance/helpfulness ratings (ranging from 4.0 to 5.0). We chose these particular divisions and to use shades of one color so that all ratings stronger than neutral (3.00) would be highlighted in blue, and the darker the blue, the stronger the level of agreement on importance/helpfulness.

Table E.1
Officer Importance Ratings, by AFSC, Grade, and Deployment Subgroups

Light Blue = Mild Importance Ratings of 3.1–3.4 | Medium Blue = Medium Importance Ratings of 3.5–3.9 | Dark Blue = High Importance Ratings of 4.0–5.0

2-Digit AFSC	Deployed Location	Grade	Gather/Interpret Info	Apply Social Etiquette—SHORT	Respect Cultural Diffs	Apply Social Etiquette	Self-Initiated Learning—SHORT	Change Behavior—SHORT	Establish Credibility	Change Behavior	Verbal/Nonverbal Comm	Manage Stress	Item: Speak Language	Negotiate with Others	Apply Regional Knowledge	Resolve Conflict	Self-Initiated Learning	Foreign Lang Skills	Influence Others	Establish Authority	# Respondents	Invited Sample	Population
All Deployed Personnel	All Deployed Personnel	All Deployed Personnel	3.4	3.3	3.4	3.3	3.1	3.1	3.1	3.0	3.1	3.1	2.9	3.0	2.9	2.8	2.9	2.7	2.7	2.5	5,374-6,015	23,020	88,304
Commander (10XX)	IRAQ/AFG	LO																			0	0	0
		HI																			3-4	18	18
	OTHER	LO																			0	0	0
		HI	3.4	3.6	3.6	3.6	3.3	3.3	3.7	3.3	3.3	2.8	3.3	3.6	3.0	2.8	3.1	3.2	3.2	2.4	8	32	32
Pilot (11XX)	IRAQ/AFG	LO	2.8	2.7	2.9	2.7	2.7	2.4	2.6	2.4	2.5	2.6	2.5	2.4	2.8	2.2	2.6	2.5	2.3	2.1	26-33	150	700
		HI	3.0	3.0	3.0	3.0	2.9	2.8	2.9	2.7	2.9	2.8	2.9	2.8	3.1	2.6	2.7	2.6	2.7	2.2	57-68	150	392
	OTHER	LO	3.0	3.0	3.0	3.0	2.7	2.7	2.7	2.6	2.6	2.5	2.7	2.7	2.7	2.4	2.6	2.5	2.2	2.1	24-31	150	2,711
		HI	3.0	3.2	3.3	3.2	2.8	2.9	2.7	2.8	2.8	2.7	2.9	2.7	2.8	2.3	2.5	2.6	2.3	1.9	47-60	150	962
Navigator (12XX)	IRAQ/AFG	LO	2.8	2.7	2.7	2.6	2.7	2.4	2.5	2.3	2.7	2.6	2.8	2.2	2.9	2.3	2.5	2.5	2.2	2.1	33-35	150	254
		HI	2.6	2.5	2.7	2.4	2.7	2.4	2.3	2.3	2.4	2.4	2.4	2.3	2.8	2.2	2.5	2.2	2.3	2.1	32-36	150	161
	OTHER	LO	3.2	3.3	3.3	3.2	3.0	3.1	3.0	2.9	3.0	3.0	2.9	3.0	2.8	2.5	2.8	2.9	2.5	2.2	28-31	150	900
		HI	3.2	3.3	3.3	3.3	3.0	3.1	2.9	2.9	2.9	2.8	2.9	2.7	2.8	2.3	2.7	2.6	2.4	1.8	60-71	150	372
Space, Missile, and Command and Control (13XX)	IRAQ/AFG	LO	3.1	3.1	3.2	3.1	3.0	2.7	3.0	2.7	3.0	3.0	3.0	2.9	3.0	2.6	2.8	2.7	2.6	2.5	29-35	150	153
		HI	3.1	3.0	3.1	2.9	3.0	2.8	2.8	2.7	2.8	2.6	3.0	2.7	3.1	2.3	2.7	2.5	2.5	2.1	16-21	70	70
	OTHER	LO	3.8	3.9	3.8	3.8	3.6	3.6	3.6	3.5	3.5	3.6	3.6	3.5	3.3	3.0	3.4	3.2	3.0	2.5	26-30	150	413
		HI	3.3	3.4	3.4	3.4	2.9	3.2	3.0	3.0	2.9	3.0	3.0	3.0	3.0	2.6	2.7	2.7	2.3	1.9	54-64	150	184
Intelligence (14XX)	IRAQ/AFG	LO	3.6	3.6	3.6	3.6	3.6	3.4	3.3	3.3	3.2	3.3	3.5	3.0	4.1	2.9	3.4	3.3	2.8	2.5	45-52	150	236
		HI	3.8	3.9	3.9	3.9	3.5	3.7	3.4	3.6	3.7	3.3	3.5	3.4	3.9	3.1	3.3	3.3	3.3	2.7	21-26	68	68
	OTHER	LO	3.3	3.3	3.5	3.3	3.0	3.0	2.8	2.9	2.7	2.9	2.6	2.8	3.0	2.3	2.7	2.5	2.2	1.9	38-46	150	427
		HI	3.3	3.5	3.5	3.4	3.1	3.3	2.9	3.1	2.9	3.2	3.2	3.0	3.1	2.5	2.9	2.9	2.4	2.0	35-36	108	108
Weather (15XX)	IRAQ/AFG	LO	3.3	3.7	3.6	3.7	3.4	3.2	3.1	3.1	3.4	3.3	3.5	3.1	3.2	2.4	3.1	3.0	2.5	2.4	5-6	37	37
		HI	2.8	2.6	2.9	2.6	2.6	2.6	2.4	2.5	2.4	2.5	2.6	2.6	2.4	2.1	2.4	2.4	2.6	1.9	6-7	15	15
	OTHER	LO	3.7	3.8	3.8	3.7	3.7	3.6	3.3	3.3	3.4	3.4	3.3	3.2	3.1	2.7	3.3	2.9	2.6	2.0	11-12	38	38
		HI																			2-3	9	9
International Affairs/FAO (16XX)	IRAQ/AFG	LO	4.4	4.2	4.3	4.1	4.3	4.1	4.3	4.0	4.1	4.1	4.5	4.0	4.2	4.0	3.9	4.0	3.8	3.2	10-13	6	6
		HI	4.0	4.0	3.9	4.0	3.9	3.8	4.0	3.7	3.9	3.6	4.0	3.7	3.8	3.4	3.5	3.4	3.7	2.8	45-48	39	39
	OTHER	LO	3.6	3.5	3.2	4.0	3.4	3.3	2.7	3.1	2.3	2.7	2.3	2.5	3.0	2.3	2.9	1.9	2.2	1.7	5-6	5	5
		HI	3.6	3.9	3.8	3.7	3.6	3.5	3.3	3.4	3.4	3.3	3.4	3.4	3.0	2.6	3.3	3.3	2.7	1.0	17-21	43	43
Logistics Commander (20XX)	IRAQ/AFG	LO																			0	0	0
		HI																			4	18	18
	OTHER	LO																			0	0	0
		HI	3.1	3.8	3.5	3.7	2.9	3.4	3.2	3.2	2.7	2.8	2.3	2.8	2.5	2.0	2.6	1.9	2.0	1.6	6-7	10	10
Aircraft Maintenance (21XX)	IRAQ/AFG	LO	3.5	3.4	3.4	3.3	3.0	3.3	3.3	3.2	3.3	3.2	3.2	3.2	3.3	3.1	2.9	2.9	3.2	2.9	34-37	150	295
		HI	3.3	3.5	3.5	3.5	3.3	3.3	3.4	3.2	3.3	3.2	3.4	3.3	3.1	3.0	3.0	2.9	3.1	2.4	49-56	133	133
	OTHER	LO	3.3	3.2	3.3	3.1	3.0	2.9	2.8	2.8	2.8	2.7	2.9	2.6	2.8	2.4	2.8	2.6	2.3	2.1	32-39	150	470
		HI	3.5	3.7	3.7	3.5	3.3	3.4	3.1	3.2	3.2	3.2	3.3	3.2	2.9	2.7	3.0	2.8	2.6	2.1	41-46	150	153

RAND *MG811-T.E.1a*

Table E.1—continued

2-Digit AFSC	Deployed Location	Grade	Gather/Interpret Info	Apply Social Etiquette—SHORT	Respect Cultural Diffs	Apply Social Etiquette	Self-Initiated Learning—SHORT	Change Behavior—SHORT	Establish Credibility	Change Behavior	Verbal/Nonverbal Comm	Manage Stress	Item: Speak Language	Negotiate with Others	Apply Regional Knowledge	Resolve Conflict	Self-Initiated Learning	Foreign Lang Skills	Influence Others	Establish Authority	# Respondents	Invited Sample	Population
Support Commander (30XX)	IRAQ/AFG	LO																			0	0	0
		HI	4.0	4.1	4.1	4.1	3.6	3.8	4.0	3.8	3.9	3.8	4.2	3.6	3.6	3.7	3.3	3.5	3.7	3.1	9-10	26	26
	OTHER	LO																			0	0	0
		HI	3.9	4.1	4.1	3.9	3.6	3.8	4.0	3.6	3.6	3.3	3.5	3.7	3.6	3.1	3.3	3.3	3.0	2.4	9-11	26	26
Security Forces (31XX)	IRAQ/AFG	LO	4.4	4.2	4.2	4.2	3.9	4.0	4.2	3.9	4.3	3.9	4.1	4.0	4.0	4.0	3.6	3.7	4.0	3.8	29-31	150	178
		HI	4.5	4.4	4.3	4.4	4.1	4.2	4.4	4.1	4.3	3.7	4.4	4.1	4.0	4.1	3.8	3.8	4.2	3.9	12	45	45
	OTHER	LO	4.4	4.3	4.1	4.2	4.1	4.1	4.2	4.0	4.2	3.9	3.8	4.1	3.9	4.0	3.9	3.1	3.6	3.1	7-9	50	50
		HI																			2	12	12
Civil Engineer (32XX)	IRAQ/AFG	LO	3.8	3.8	3.7	3.8	3.4	3.5	3.7	3.3	3.8	3.5	3.6	3.8	3.4	3.1	3.2	3.2	3.4	2.9	31-41	150	218
		HI	3.6	3.7	3.7	3.7	3.4	3.4	3.6	3.3	3.5	3.4	3.3	3.5	3.4	3.0	3.1	2.9	3.2	2.5	39-43	98	98
	OTHER	LO	3.9	4.0	4.0	3.8	3.8	3.7	3.8	3.5	3.9	3.3	3.7	4.0	3.4	3.2	3.5	3.4	3.3	2.7	24-26	139	139
		HI	3.6	3.8	3.7	3.8	3.4	3.4	3.3	3.3	3.4	3.2	3.1	3.5	3.1	2.7	3.1	2.8	2.6	2.3	13-17	48	48
Comm-Information Systems (33XX)	IRAQ/AFG	LO	3.4	3.4	3.5	3.4	3.3	3.1	3.2	3.1	3.2	3.2	3.3	3.1	3.2	2.7	3.0	2.9	2.9	2.3	37-41	150	317
		HI	3.3	3.3	3.4	3.4	3.2	3.2	3.3	3.1	3.1	3.1	3.1	3.2	3.1	2.9	2.9	2.9	3.0	2.4	39-46	91	91
	OTHER	LO	3.2	3.5	3.3	3.5	3.3	3.1	3.0	3.0	3.0	3.1	3.2	3.1	3.1	2.6	3.1	2.9	2.6	2.4	25-29	150	397
		HI	3.3	3.5	3.6	3.4	3.1	3.3	3.0	3.0	3.0	3.1	2.9	3.0	2.9	2.5	2.8	2.6	2.4	1.9	27-52	94	94
Services (34XX)	IRAQ/AFG	LO	3.8	4.0	3.9	4.1	3.6	3.5	3.6	3.4	3.5	3.2	3.7	3.5	3.4	2.9	3.3	2.9	3.0	2.6	17-19	45	45
		HI																			2-3	15	15
	OTHER	LO	3.7	4.1	3.9	4.0	3.7	3.6	3.5	3.4	3.4	3.5	3.6	3.6	3.2	2.6	3.4	3.0	2.7	2.3	19-20	87	87
		HI	3.9	3.9	3.8	3.7	3.2	3.7	3.5	3.5	3.9	3.7	3.8	3.7	3.5	3.3	3.0	3.0	3.3	2.8	7-8	18	18
Public Affairs (35XX)	IRAQ/AFG	LO	4.1	4.6	4.1	4.5	4.0	4.2	3.7	4.0	3.9	3.6	3.5	3.5	3.7	3.3	3.5	3.7	3.4	2.4	5-6	32	32
		HI	3.2	3.1	3.5	3.0	3.5	3.1	3.3	3.0	3.1	2.9	3.0	3.1	3.6	3.0	3.3	3.5	2.9	2.0	6-7	11	11
	OTHER	LO	3.8	4.1	4.1	3.9	3.7	3.7	3.8	3.4	3.9	3.2	3.5	3.9	3.5	3.4	3.3	3.6	3.3	2.0	8-11	40	40
		HI																			3-5	25	25
Personnel (36XX)	IRAQ/AFG	LO																			0	26	26
		HI																			0	15	15
	OTHER	LO																			0	41	41
		HI																			0	9	9
Manpower-Personnel (37XX)	IRAQ/AFG	LO	3.0	3.2	3.1	3.2	3.0	2.9	2.7	2.7	2.6	2.8	3.1	2.6	2.9	2.3	2.9	2.7	2.2	2.1	15-19	51	51
		HI	3.6	3.8	3.8	3.7	3.4	3.6	3.4	3.4	3.5	3.4	3.5	3.3	3.2	3.3	3.2	2.9	3.2	2.7	22-23	36	36
	OTHER	LO	3.7	3.8	3.8	3.7	3.6	3.5	3.2	3.3	3.4	3.3	3.6	3.3	3.2	2.8	3.3	3.2	2.7	2.2	40-42	83	83
		HI	3.6	3.9	3.7	3.8	3.4	3.6	3.4	3.5	3.4	3.4	3.2	3.3	3.4	2.9	3.0	3.0	2.8	2.1	13-14	31	31
Manpower (38XX)	IRAQ/AFG	LO																			0	3	3
		HI																			0	1	1
	OTHER	LO																			0	13	13
		HI																			0	3	3
Medical Commander (40XX)	IRAQ/AFG	LO																			0	0	0
		HI																			4-5	7	7
	OTHER	LO																			0	0	0
		HI	3.5	3.9	4.1	3.8	3.2	3.6	3.5	3.4	3.2	3.4	3.1	3.3	3.5	2.9	2.9	2.7	3.0	1.9	6-7	14	14
Health Services Administrator (41XX)	IRAQ/AFG	LO	2.8	2.9	3.1	2.9	2.9	3.0	2.7	3.1	2.6	2.4	2.6	2.2	2.5	2.1	2.7	2.2	2.4	2.1	8-10	40	40
		HI	3.3	3.5	3.4	3.4	3.4	3.2	3.4	3.3	3.2	3.3	3.3	3.0	2.9	2.8	3.2	2.9	3.0	2.4	14-15	44	44
	OTHER	LO	3.5	3.4	3.6	3.3	3.6	3.2	2.9	3.0	3.3	3.3	3.8	3.1	2.8	2.5	3.4	3.2	2.4	2.4	7-8	44	44
		HI	3.7	4.2	4.0	4.0	3.5	3.6	3.3	3.4	3.7	3.2	3.2	3.4	3.4	2.7	3.1	3.0	2.8	2.0	10-14	37	37

RAND *MG811-T.E.1b*

Table E.1—continued

2-Digit AFSC	Deployed Location	Grade	Gather/Interpret Info	Apply Social Etiquette—SHORT	Respect Cultural Diffs	Apply Social Etiquette	Self-Initiated Learning—SHORT	Change Behavior—SHORT	Establish Credibility	Change Behavior	Verbal/Nonverbal Comm	Manage Stress	Item: Speak Language	Negotiate with Others	Apply Regional Knowledge	Resolve Conflict	Self-Initiated Learning	Foreign Lang Skills	Influence Others	Establish Authority	# Respondents	Invited Sample	Population
Biomedical Clinician (42XX)	IRAQ/AFG	LO	3.9	3.8	3.9	3.9	3.6	3.5	3.7	3.4	3.5	3.5	3.3	3.2	3.4	2.9	3.3	2.8	3.1	2.3	20-25	70	70
		HI	3.0	2.9	3.1	2.9	2.8	2.6	2.6	2.6	2.9	2.5	3.3	2.1	3.0	2.4	2.5	2.1	2.6	2.0	8-11	34	34
	OTHER	LO																			2-3	48	48
		HI																			3-4	17	17
Biomedical Specialists (43XX)	IRAQ/AFG	LO	3.5	3.3	3.4	3.4	3.5	3.1	3.4	3.0	3.5	3.4	3.3	3.3	3.4	3.0	3.3	2.8	3.2	2.9	7-8	32	32
		HI	2.8	2.9	2.9	2.9	2.7	2.6	2.6	2.6	2.7	2.7	2.8	2.5	2.5	2.3	2.5	2.3	2.4	2.0	18-21	49	49
	OTHER	LO	3.8	3.9	3.8	3.8	3.6	3.6	3.5	3.6	3.4	3.6	3.2	3.2	3.2	3.0	3.3	3.0	2.7	2.2	11-13	34	34
		HI	3.4	3.7	3.5	3.7	3.4	3.5	3.1	3.3	3.2	3.2	3.0	3.2	3.1	2.6	3.1	2.8	2.6	2.1	20-22	42	42
Physician (44XX)	IRAQ/AFG	LO																			3-6	46	46
		HI	3.4	3.6	3.8	3.7	3.2	3.2	3.4	3.1	3.2	3.2	3.2	2.8	3.0	2.5	3.0	2.5	2.9	2.1	29-33	120	120
	OTHER	LO																			1-3	48	48
		HI	3.5	3.8	3.9	3.7	3.5	3.4	3.3	3.2	3.3	3.1	3.4	3.3	3.1	2.7	3.2	3.0	2.8	2.2	21-23	77	77
Surgery (45XX)	IRAQ/AFG	LO																			1	16	16
		HI	3.9	3.9	4.0	3.9	3.8	3.7	3.9	3.6	4.0	3.8	3.8	3.4	3.4	3.3	3.4	3.5	3.4	2.8	16-18	113	113
	OTHER	LO																			0	4	4
		HI	3.5	3.5	3.7	3.2	3.3	2.9	3.3	2.8	3.0	2.8	2.9	3.0	2.9	2.2	3.0	2.7	2.4	1.9	7-8	38	38
Nurse (46XX)	IRAQ/AFG	LO	3.4	3.5	3.6	3.5	3.2	3.2	3.3	3.1	3.4	3.4	3.6	2.9	3.0	2.8	3.0	2.9	2.9	2.6	41-47	150	258
		HI	3.4	3.5	3.7	3.5	3.4	3.2	3.2	3.1	3.4	3.4	3.3	2.8	3.2	2.7	3.1	2.8	2.7	2.3	41-47	99	99
	OTHER	LO	3.6	3.7	3.8	3.7	3.5	3.5	3.3	3.3	3.4	3.4	3.6	3.2	3.2	2.9	3.3	3.2	2.8	2.5	44-49	150	269
		HI	3.5	3.7	3.7	3.7	3.3	3.4	3.1	3.2	3.1	3.2	3.0	3.0	3.0	2.6	3.0	2.8	2.5	2.1	25-33	94	94
Dental (47XX)	IRAQ/AFG	LO																			0	4	4
		HI																			3-4	12	12
	OTHER	LO																			1	8	8
		HI	3.6	3.8	3.7	3.7	3.5	3.6	3.5	3.4	3.4	3.5	3.5	3.3	3.2	3.0	3.3	3.1	3.1	2.6	10-12	24	24
Aerospace Medicine (48XX)	IRAQ/AFG	LO																			3-5	26	26
		HI	3.3	3.3	3.4	3.3	3.1	2.9	3.3	2.8	3.3	3.1	3.2	2.9	3.0	2.4	2.8	2.6	2.8	2.0	13-16	51	51
	OTHER	LO	3.2	3.4	3.7	3.6	3.0	3.2	3.3	3.0	3.1	3.4	2.0	2.9	3.0	2.2	2.8	2.1	2.5	1.9	5-6	57	57
		HI	3.4	3.8	3.6	3.8	3.3	3.5	3.3	3.3	3.3	3.2	3.3	3.3	3.1	2.6	3.0	2.9	2.6	2.0	27-30	81	81
Law (51XX)	IRAQ/AFG	LO	3.3	3.6	3.4	3.6	3.0	3.5	3.5	3.4	3.3	2.8	3.0	3.1	3.6	3.0	2.8	2.6	3.1	3.2	7-9	56	56
		HI	3.4	3.5	3.4	3.4	3.2	3.2	3.2	3.1	3.0	2.9	3.1	3.2	3.4	2.9	3.0	2.9	3.0	2.6	20-21	43	43
	OTHER	LO																			2-3	37	37
		HI	3.4	3.5	3.6	3.4	3.4	3.1	3.0	3.0	3.2	3.1	3.6	3.2	3.3	3.0	3.2	3.0	2.8	2.4	15-16	49	49
Chaplain (52XX)	IRAQ/AFG	LO	3.6	3.8	3.9	3.7	3.7	3.4	3.4	3.4	3.7	3.3	3.7	3.2	3.4	2.9	3.3	3.1	2.8	1.9	12-15	43	43
		HI	3.5	3.7	3.8	3.6	3.3	3.4	3.3	3.2	3.2	3.2	3.4	2.9	3.5	2.8	3.0	2.9	2.8	2.2	16-10	49	49
	OTHER	LO	3.8	4.0	3.9	4.0	3.8	3.8	3.6	3.5	3.5	3.7	3.8	3.4	3.7	2.9	3.5	3.4	2.7	2.1	10-11	43	43
		HI	3.6	3.9	3.7	3.8	3.6	3.6	3.4	3.4	3.4	3.3	3.5	3.5	3.6	2.6	3.4	3.1	2.8	2.2	9-12	25	25
Program Director (60XX)	IRAQ/AFG	LO																			0	0	0
		HI																			0	1	1
	OTHER	LO																			0	0	0
		HI																			0	0	0

RAND *MG811-T.E.1c*

Table E.1—continued

2-Digit AFSC	Deployed Location	Grade	Gather/Interpret Info	Apply Social Etiquette—SHORT	Respect Cultural Diffs	Apply Social Etiquette	Self-Initiated Learning—SHORT	Change Behavior—SHORT	Establish Credibility	Change Behavior	Verbal/Nonverbal Comm	Manage Stress	Item: Speak Language	Negotiate with Others	Apply Regional Knowledge	Resolve Conflict	Self-Initiated Learning	Foreign Lang Skills	Influence Others	Establish Authority	# Respondents	Invited Sample	Population
Scientific/ Research (61XX)	IRAQ/AFG	LO																			3	9	9
		HI																			3-4	4	4
	OTHER	LO																			4-5	15	15
		HI	3.2	3.4	3.3	3.3	2.9	3.4	2.7	3.3	2.6	3.1	2.7	2.8	2.6	2.0	2.6	2.5	1.9	1.7	6-7	16	16
Developmental Engineering (62XX)	IRAQ/AFG	LO	3.3	3.5	3.6	3.4	3.2	3.1	3.2	3.2	3.1	3.2	3.3	2.9	3.2	2.3	3.0	2.7	2.7	2.0	6	21	21
		HI																			2	10	10
	OTHER	LO	2.5	2.9	3.3	2.9	2.5	2.5	2.2	2.4	2.4	2.3	2.0	2.2	2.1	1.8	2.3	2.0	1.8	1.7	0-7	31	31
		HI																			1	5	5
Acquisition (63XX)	IRAQ/AFG	LO	4.2	4.0	4.0	3.9	3.8	3.9	3.8	3.6	4.1	3.7	4.0	3.7	3.9	3.1	3.3	3.5	3.5	2.9	7-8	33	33
		HI	3.5	3.8	3.6	3.8	3.6	3.5	3.7	3.4	3.4	3.1	3.7	3.4	3.6	2.8	3.3	3.1	3.3	2.6	9	42	42
	OTHER	LO																			2-3	31	31
		HI																			3-4	17	17
Contracting (64XX)	IRAQ/AFG	LO	4.2	4.3	4.2	4.2	3.9	4.0	4.1	3.8	4.0	3.8	3.7	4.3	3.7	3.5	3.6	3.2	3.4	3.0	31-34	111	111
		HI	3.8	4.1	4.1	4.1	3.8	3.5	3.9	3.5	3.8	3.3	3.6	4.2	3.8	3.4	3.4	3.2	3.5	3.0	19-21	52	52
	OTHER	LO	4.0	4.3	4.3	4.3	4.0	3.9	3.9	3.7	4.0	3.5	4.0	4.5	3.4	3.1	3.6	3.5	3.2	2.6	20-22	69	69
		HI																			3-4	4	4
Finance (65XX)	IRAQ/AFG	LO	3.7	4.0	3.8	4.0	3.7	3.3	3.5	3.1	3.5	3.7	3.8	3.3	3.4	2.7	3.4	2.9	3.1	2.4	7-9	43	43
		HI																			4-6	37	37
	OTHER	LO																			4-5	22	22
		HI	3.9	4.0	4.1	3.9	4.0	3.9	3.9	3.8	3.7	3.4	3.4	3.6	3.6	3.2	3.6	2.8	3.3	2.7	8-9	17	17
Special Investigations (71XX)	IRAQ/AFG	LO																			3-4	30	30
		HI																			3-4	14	14
	OTHER	LO																			3-4	17	17
		HI																			1	4	4
Commander—USAFA Cadet Squadron (80XX)	IRAQ/AFG	LO																			0	0	0
		HI																			0	0	0
	OTHER	LO																			0	0	0
		HI																			0	1	1
Instructor (81XX)	IRAQ/AFG	LO																			1	4	4
		HI																			2-3	8	8
	OTHER	LO																			0	6	6
		HI																			0	8	8
Academic Program Manager (82XX)	IRAQ/AFG	LO																			0	1	1
		HI																			0	3	3
	OTHER	LO																			0	1	1
		HI																			0	3	3
Recruiting Service (83XX)	IRAQ/AFG	LO																			0	4	4
		HI																			0	0	0
	OTHER	LO																			0	2	2
		HI																			0	0	0

RAND *MG811-T.E.1d*

Table E.1—continued

2-Digit AFSC	Deployed Location	Grade	Gather/Interpret Info	Apply Social Etiquette—SHORT	Respect Cultural Diffs	Apply Social Etiquette	Self-Initiated Learning—SHORT	Change Behavior—SHORT	Establish Credibility	Change Behavior	Verbal/Nonverbal Comm	Manage Stress	Item: Speak Language	Negotiate with Others	Apply Regional Knowledge	Resolve Conflict	Self-Initiated Learning	Foreign Lang Skills	Influence Others	Establish Authority	# Respondents	Invited Sample	Population
Historian (84XX)	IRAQ/AFG	LO																			0	0	0
		HI																			0	0	0
	OTHER	LO																			0	0	0
		HI																			0	0	0
USAF Honor Guard (85XX)	IRAQ/AFG	LO																			0	0	0
		HI																			0	0	0
	OTHER	LO																			0	0	0
		HI																			0	1	1
Command and Control (86XX)	IRAQ/AFG	LO																			0	2	2
		HI																			3-4	3	3
	OTHER	LO																			2-3	3	3
		HI																			1	5	5
Inspector General (87XX)	IRAQ/AFG	LO																			0	0	0
		HI																			0	0	0
	OTHER	LO																			0	0	0
		HI																			0	1	1
Protocol/Aide (88XX)	IRAQ/AFG	LO																			2	1	1
		HI																			1	0	0
	OTHER	LO																			3	0	0
		HI																			0	0	0
General Officer (90XX)	IRAQ/AFG	LO																			0	0	0
		HI																			2-3	10	10
	OTHER	LO																			0	0	0
		HI																			0	11	11
Commander (91XX)	IRAQ/AFG	LO																			0	0	0
		HI																			3-4	13	13
	OTHER	LO																			0	0	0
		HI	3.5	3.8	3.7	3.5	3.1	3.5	3.5	3.3	3.1	3.1	2.9	3.4	3.3	2.7	2.8	2.6	2.7	2.2	9	18	18
Student (92XX)	IRAQ/AFG	LO																			0	22	22
		HI																			0	7	7
	OTHER	LO																			1-2	65	65
		HI																			0	12	12
Patient (93XX)	IRAQ/AFG	LO																			0	0	0
		HI																			0	0	0
	OTHER	LO																			0	0	0
		HI																			0	0	0
Nuclear Weapons Custodian (94XX)	IRAQ/AFG	LO																			0	0	0
		HI																			0	0	0
	OTHER	LO																			0	0	0
		HI																			0	0	0

RAND *MG811-T.E.1e*

Table E.1—continued

2-Digit AFSC	Deployed Location	Grade	Gather/Interpret Info	Apply Social Etiquette—SHORT	Respect Cultural Diffs	Apply Social Etiquette	Self-Initiated Learning—SHORT	Change Behavior—SHORT	Establish Credibility	Change Behavior	Verbal/Nonverbal Comm	Manage Stress	Item: Speak Language	Negotiate with Others	Apply Regional Knowledge	Resolve Conflict	Self-Initiated Learning	Foreign Lang Skills	Influence Others	Establish Authority	# Respondents	Invited Sample	Population
Non-extended Active Duty (95XX)	IRAQ/AFG	LO																			0	0	0
		HI																			0	0	0
	OTHER	LO																			0	0	0
		HI																			0	1	1
Unclassified Officer (96XX)	IRAQ/AFG	LO																			0	0	0
		HI																			0	0	0
	OTHER	LO																			1	2	2
		HI																			0	0	0
Executive Officer above Wing Level (97XX)	IRAQ/AFG	LO																			1	3	3
		HI																			2-3	12	12
	OTHER	LO																			1-2	6	6
		HI																			0-1	7	7

RAND *MG811-T.E.1f*

Table E.2
Enlisted Importance Ratings, by AFSC, Grade, and Deployment Subgroups

Light Blue = Mild Importance Ratings of 3.1–3.4 | Medium Blue = Medium Importance Ratings of 3.5–3.9 | Dark Blue = High Importance Ratings of 4.0–5.0

2-Digit AFSC	Deployed Location	Grade	Gather/Interpret Info	Apply Social Etiquette—SHORT	Respect Cultural Diffs	Apply Social Etiquette	Self-Initiated Learning—SHORT	Change Behavior—SHORT	Establish Credibility	Change Behavior	Verbal/Nonverbal Comm	Manage Stress	Item: Speak Language	Negotiate with Others	Apply Regional Knowledge	Resolve Conflict	Self-Initiated Learning	Foreign Lang Skills	Influence Others	Establish Authority	# Respondents	Invited Sample	Population
All Deployed Personnel	All Deployed Personnel	All Deployed Personnel	3.4	3.3	3.4	3.3	3.1	3.1	3.1	3.0	3.1	3.1	2.9	3.0	2.9	2.8	2.9	2.7	2.7	2.5	5,374-6,015	23,020	88,304
Aircrew Operations (1AXXX)	IRAQ/AFG	LO	2.5	2.4	2.5	2.3	2.3	2.5	2.5	2.3	2.4	2.3	2.2	2.3	2.3	2.3	2.3	2.2	2.3	2.2	12-14	150	218
		HI	3.0	2.7	2.9	2.7	2.6	2.5	2.7	2.5	2.4	2.7	2.5	2.5	2.6	2.4	2.5	2.5	2.4	2.2	30-34	150	633
	OTHER	LO	3.5	3.7	3.7	3.6	3.2	3.3	3.3	3.2	3.1	3.3	3.2	3.2	3.0	2.6	3.0	3.1	2.6	2.3	19-23	150	1,248
		HI	3.0	3.0	3.0	2.9	2.7	2.8	2.7	2.7	2.8	2.7	2.6	2.6	2.6	2.4	2.6	2.6	2.3	2.1	34-40	150	1,431
Command & Control Systems Operations (1CXXX)	IRAQ/AFG	LO	3.0	2.9	3.1	2.9	2.8	2.8	2.6	2.8	2.7	2.9	2.3	2.4	2.7	2.6	2.6	2.4	2.3	2.4	13-17	150	859
		HI	3.0	3.0	3.0	3.0	2.8	2.8	2.8	2.8	3.0	2.9	2.7	2.8	2.9	2.7	2.6	2.4	2.6	2.5	45-52	150	983
	OTHER	LO	3.3	3.5	3.5	3.5	3.2	3.2	3.1	3.1	3.2	3.1	3.0	3.1	3.0	2.7	3.1	2.7	2.7	2.6	16-20	150	554
		HI	3.4	3.6	3.6	3.5	3.2	3.2	3.0	3.0	3.0	3.1	2.9	2.9	3.0	2.7	3.0	2.8	2.5	2.2	53-65	150	676
Intelligence (1NXXX)	IRAQ/AFG	LO	3.8	4.0	3.8	3.9	3.5	3.7	3.6	3.6	3.6	3.4	3.4	3.5	3.4	3.6	3.3	3.0	3.4	3.0	11-15	150	290
		HI	3.4	3.4	3.4	3.3	3.4	3.3	3.1	3.1	3.3	3.3	3.2	3.0	3.5	2.8	3.1	3.1	2.9	2.5	49-54	150	313
	OTHER	LO	3.3	3.5	3.5	3.5	3.2	3.3	2.9	3.1	3.2	3.1	2.5	3.1	3.2	2.4	3.0	2.4	2.4	2.1	17-24	150	436
		HI	3.6	3.8	3.7	3.7	3.4	3.6	3.2	3.4	3.3	3.2	3.1	3.2	3.2	2.9	3.2	3.0	2.7	2.3	43-48	150	437
Safety (1SXXX)	IRAQ/AFG	LO																			0	0	0
		HI	3.0	3.1	3.1	3.0	3.0	3.0	2.8	2.9	2.8	2.7	2.7	2.5	2.7	2.6	2.8	2.6	2.5	2.2	10-13	39	39
	OTHER	LO																			0	1	1
		HI	3.8	3.5	3.4	3.5	3.3	3.2	3.1	2.9	3.3	3.0	3.4	3.1	2.9	3.0	3.2	2.9	2.8	2.6	10-16	41	41
Aircrew Protection (1TXXX)	IRAQ/AFG	LO	3.1	3.0	3.0	3.1	2.9	3.1	3.0	3.0	3.0	3.0	3.0	2.8	2.8	2.8	2.9	2.8	2.8	2.6	6-9	115	115
		HI	3.6	3.6	3.5	3.5	3.5	3.4	3.1	3.3	3.3	3.3	3.4	3.1	3.1	3.0	3.3	3.0	3.0	2.8	22-28	118	118
	OTHER	LO	3.6	3.8	3.6	3.8	3.6	3.5	3.5	3.3	3.1	3.6	3.3	3.3	3.4	3.0	3.3	2.9	2.8	2.4	8-13	150	276
		HI	3.6	3.5	3.6	3.5	3.3	3.5	3.2	3.3	3.3	3.3	3.4	3.0	3.1	2.8	3.1	3.2	2.6	2.3	39-47	150	290
Weather (1WXXX)	IRAQ/AFG	LO																			4-6	68	68
		HI	2.9	2.9	3.0	2.9	2.9	2.8	2.7	2.6	2.7	2.8	2.7	2.6	2.6	2.4	2.8	2.5	2.4	2.3	49-57	150	217
	OTHER	LO																			3-4	54	54
		HI	3.3	3.2	3.4	3.2	2.9	2.9	2.9	2.8	2.8	2.9	2.6	2.7	2.9	2.4	2.6	2.3	2.2	1.9	31-41	127	127
Manned Aerospace Maintenance (2AXXX)	IRAQ/AFG	LO	3.5	3.2	3.4	3.3	3.2	3.2	3.1	3.2	2.9	3.2	3.0	3.0	3.1	3.0	3.0	3.0	2.8	2.5	8-11	150	2,505
		HI	3.1	3.0	3.2	3.0	2.8	2.9	2.8	2.8	2.8	2.9	2.6	2.6	2.7	2.7	2.6	2.6	2.6	2.5	48-59	150	2,396
	OTHER	LO	3.3	3.1	3.3	3.2	2.8	3.2	2.0	2.0	2.9	3.0	2.1	2.7	2.6	2.5	2.7	2.0	2.2	2.3	8-13	150	5,727
		HI	3.1	3.1	3.1	3.0	2.8	2.9	2.8	2.8	2.8	2.7	2.6	2.7	2.6	2.6	2.6	2.6	2.5	2.3	47-54	150	5,920
Communications & Electronics (2EXXX)	IRAQ/AFG	LO	3.3	3.1	3.1	3.1	2.7	3.1	2.9	2.9	2.9	3.0	2.5	2.8	2.8	2.8	2.7	2.6	2.6	2.8	18-25	150	705
		HI	2.7	2.7	2.8	2.7	2.5	2.6	2.6	2.5	2.5	2.5	2.5	2.3	2.5	2.3	2.4	2.3	2.2	2.1	66-75	150	596
	OTHER	LO	3.5	3.5	3.6	3.5	3.2	3.3	3.3	3.0	3.4	3.2	3.0	3.2	3.2	2.9	3.1	3.0	2.9	2.7	21-29	150	654
		HI	3.3	3.4	3.4	3.3	2.8	3.2	2.9	3.0	2.9	3.0	2.7	2.8	2.7	2.5	2.7	2.5	2.2	2.1	49-57	150	689
Fuels (2FXXX)	IRAQ/AFG	LO	3.2	3.2	3.2	3.2	2.8	3.1	3.1	3.0	3.1	3.0	2.7	3.1	2.9	3.0	2.8	2.9	3.0	3.1	18-23	150	209
		HI	3.1	2.9	2.9	2.9	2.7	2.9	2.8	2.8	2.8	2.7	2.7	2.6	2.6	2.6	2.5	2.5	2.6	2.5	46-57	150	264
	OTHER	LO	3.8	3.6	3.7	3.6	3.1	3.6	3.2	3.3	3.5	3.3	2.8	3.5	3.1	3.0	2.8	2.9	3.0	2.8	11-15	150	668
		HI	3.5	3.5	3.4	3.4	3.1	3.2	3.2	3.1	3.3	3.1	3.3	3.2	3.0	2.9	3.0	2.9	2.9	2.7	59-67	150	709

RAND *MG811-T.E.2a*

Table E.2—continued

2-Digit AFSC	Deployed Location	Grade	Gather/Interpret Info	Apply Social Etiquette—SHORT	Respect Cultural Diffs	Apply Social Etiquette	Self-Initiated Learning—SHORT	Change Behavior—SHORT	Establish Credibility	Change Behavior	Verbal/Nonverbal Comm	Manage Stress	Item: Speak Language	Negotiate with Others	Apply Regional Knowledge	Resolve Conflict	Self-Initiated Learning	Foreign Lang Skills	Influence Others	Establish Authority	# Respondents	Invited Sample	Population
Logistics Plans (2GXXX)	IRAQ/AFG	LO																			1-3	22	22
		HI	2.7	2.6	2.7	2.6	2.3	2.6	2.4	2.5	2.5	2.6	2.3	2.5	2.3	2.2	2.3	2.2	2.2	2.1	22-26	87	87
	OTHER	LO	3.6	3.5	3.3	3.4	3.4	3.4	3.2	3.3	3.3	3.5	3.6	3.2	3.5	3.0	3.3	3.6	3.0	2.6	6-7	37	37
		HI	3.4	3.3	3.3	3.3	3.1	3.1	2.9	3.0	3.0	3.0	2.8	2.7	2.8	2.6	2.9	2.6	2.4	2.3	31-35	119	119
Missile & Space Systems Maintenance (2MXXX)	IRAQ/AFG	LO																			1-4	59	59
		HI																			4-5	22	22
	OTHER	LO																			1-2	39	39
		HI																			1	7	7
Precision Measurement Equipment Lab (2PXXX)	IRAQ/AFG	LO																			0	5	5
		HI																			2	5	5
	OTHER	LO																			2-5	42	42
		HI	3.4	3.4	3.4	3.4	3.0	3.3	2.9	3.2	3.0	3.0	2.6	2.9	2.8	2.3	2.7	2.5	2.1	2.0	24-29	61	61
Maintenance Management Systems (2RXXX)	IRAQ/AFG	LO																			3-4	42	42
		HI	3.7	3.7	3.9	3.6	3.3	3.5	3.2	3.3	3.3	3.6	3.3	3.3	3.5	3.2	3.1	2.9	2.9	2.7	8-10	35	35
	OTHER	LO																			4-7	92	92
		HI	3.3	3.3	3.3	3.4	2.9	3.3	2.8	3.0	2.9	2.9	2.4	2.7	2.8	2.5	2.7	2.5	2.3	2.0	21-30	100	100
Supply (2SXXX)	IRAQ/AFG	LO	3.3	2.9	3.1	2.8	2.6	2.9	2.6	2.6	2.7	3.0	2.4	2.6	2.4	2.7	2.6	2.5	2.2	2.4	8-11	150	374
		HI	3.2	3.2	3.3	3.2	2.8	3.1	3.1	2.9	3.0	3.1	2.8	2.9	2.7	2.9	2.7	2.7	2.8	2.7	36-44	150	643
	OTHER	LO	3.5	3.3	3.4	3.4	3.5	3.3	3.2	3.3	3.1	3.2	3.3	3.1	3.1	3.0	3.3	3.2	3.0	2.8	11-13	150	650
		HI	3.4	3.5	3.5	3.4	3.1	3.3	3.2	3.2	3.1	3.1	3.0	3.1	3.2	3.0	3.0	2.8	2.7	2.5	32-38	150	1,067
Transportation & Vehicle Maintenance (2TXXX)	IRAQ/AFG	LO	3.4	3.2	3.2	3.2	3.0	3.0	3.0	2.9	3.0	3.0	3.0	2.9	3.1	3.0	2.9	2.8	2.8	2.8	17-23	150	1,232
		HI	3.5	3.4	3.3	3.4	3.1	3.2	3.3	3.1	3.2	3.1	3.0	3.1	3.0	3.0	2.9	2.7	2.9	2.7	55-68	150	1,219
	OTHER	LO	3.3	3.1	3.2	3.1	2.6	2.8	2.9	2.8	3.1	3.1	2.6	2.8	3.0	2.8	2.5	2.5	2.8	2.8	11-14	150	1,207
		HI	3.5	3.6	3.5	3.5	3.1	3.4	3.3	3.2	3.2	3.0	2.9	3.1	2.9	2.8	2.9	2.7	2.7	2.6	40-44	150	1,275
Munitions & Weapons (2WXXX)	IRAQ/AFG	LO	3.0	3.0	2.9	2.9	2.5	2.8	2.8	2.6	2.9	2.9	2.6	2.8	2.7	2.7	2.4	2.4	2.6	2.6	19-22	150	768
		HI	2.7	2.6	2.6	2.6	2.5	2.6	2.5	2.5	2.4	2.6	2.5	2.4	2.4	2.4	2.5	2.4	2.3	2.3	42-47	150	714
	OTHER	LO	3.6	3.5	3.5	3.4	3.3	3.2	3.2	3.1	3.2	3.3	2.9	3.0	2.9	2.9	3.1	2.8	2.8	2.8	17-22	150	980
		HI	3.0	2.9	2.9	2.9	2.8	2.9	2.7	2.8	2.8	2.8	2.8	2.6	2.6	2.5	2.6	2.6	2.4	2.4	46-53	150	905
Information Management (3AXXX)	IRAQ/AFG	LO	3.5	3.3	3.3	3.3	3.3	3.3	3.2	3.3	3.3	3.2	2.7	3.1	3.1	3.1	3.2	3.0	3.2	3.1	17-19	150	220
		HI	3.2	3.3	3.3	3.2	2.9	3.1	2.9	3.0	3.1	3.2	2.8	2.7	2.9	2.8	2.8	2.6	2.7	2.7	51-63	150	539
	OTHER	LO	3.5	3.4	3.3	3.4	3.3	3.3	3.2	3.2	3.1	3.1	3.0	3.2	3.1	2.9	3.1	3.1	2.8	2.7	15-19	150	329
		HI	3.4	3.6	3.6	3.5	3.0	3.2	3.0	3.1	2.9	3.0	2.9	2.9	2.8	2.5	2.8	2.8	2.4	2.1	43-55	150	756
Communications & Computer Systems (3CXXX)	IRAQ/AFG	LO	3.4	3.1	3.3	3.1	3.1	3.1	3.1	2.8	2.9	3.2	2.7	2.9	3.1	2.9	2.9	2.7	2.8	2.8	17-21	150	405
		HI	3.1	3.0	3.1	3.0	2.9	2.9	2.8	2.7	2.8	2.9	2.7	2.7	2.7	2.5	2.7	2.5	2.4	2.3	39-45	150	627
	OTHER	LO	3.2	3.1	3.2	3.1	2.8	3.0	2.8	3.0	2.9	3.1	2.5	2.9	2.8	2.8	2.7	2.6	2.5	2.4	11-15	150	617
		HI	3.2	3.4	3.3	3.4	2.9	3.2	2.9	3.0	2.9	2.9	2.7	2.9	2.6	2.4	2.7	2.6	2.4	2.1	41-48	150	1,026
Civil Engineering (3EXXX)	IRAQ/AFG	LO	3.2	2.9	3.1	3.0	2.9	2.8	2.9	2.8	2.7	3.0	2.7	2.8	2.7	2.7	2.8	2.5	2.7	2.9	15-18	150	1,839
		HI	3.4	3.2	3.2	3.1	3.0	3.0	3.0	2.9	3.2	3.0	3.0	3.0	2.9	2.8	2.8	2.8	2.9	2.9	55-63	150	1,826
	OTHER	LO	3.4	3.5	3.4	3.6	3.1	3.4	2.8	3.4	3.1	3.0	2.9	3.2	3.0	2.7	3.0	2.9	2.9	2.7	11-15	150	2,039
		HI	3.5	3.5	3.6	3.5	3.2	3.4	3.2	3.3	3.3	3.2	3.1	3.2	3.0	2.9	3.0	2.8	2.8	2.6	36-39	150	1,919
Historian (3HXXX)	IRAQ/AFG	LO																			0	0	0
		HI																			1-2	5	5
	OTHER	LO																			0	0	0
		HI																			3	7	7

RAND *MG811-T.E.2b*

Table E.2—continued

2-Digit AFSC	Deployed Location	Grade	Gather/Interpret Info	Apply Social Etiquette—SHORT	Respect Cultural Diffs	Apply Social Etiquette	Self-Initiated Learning—SHORT	Change Behavior—SHORT	Establish Credibility	Change Behavior	Verbal/Nonverbal Comm	Manage Stress	Item: Speak Language	Negotiate with Others	Apply Regional Knowledge	Resolve Conflict	Self-Initiated Learning	Foreign Lang Skills	Influence Others	Establish Authority	# Respondents	Invited Sample	Population
Services (3MXXX)	IRAQ/AFG	LO	3.8	3.7	3.9	3.7	3.5	3.4	3.6	3.5	3.5	3.5	3.2	3.3	3.5	3.2	3.3	3.0	3.2	3.0	19-24	150	208
		HI	3.3	3.3	3.3	3.2	3.0	3.1	3.1	2.9	3.1	3.1	2.9	3.1	2.7	2.9	2.8	2.7	2.8	2.7	40-48	150	268
	OTHER	LO	3.5	3.6	3.6	3.6	3.2	3.3	3.3	3.1	3.3	3.0	3.0	3.1	2.9	3.0	3.0	2.7	2.8	2.8	9-15	150	703
		HI	3.6	3.6	3.5	3.5	3.2	3.4	3.2	3.2	3.3	3.2	2.9	3.3	3.0	3.0	3.0	2.8	2.9	2.7	44-51	150	762
Public Affairs (3NXXX)	IRAQ/AFG																				2-4	33	33
		HI	3.9	4.0	4.0	4.0	3.7	3.9	3.8	3.7	3.7	3.7	3.5	3.4	3.6	2.9	3.4	3.3	3.0	2.4	28-33	78	78
	OTHER	LO																			2-3	20	20
		HI	3.8	3.9	3.9	3.9	3.6	3.6	3.5	3.4	3.5	3.4	3.4	3.4	3.5	3.0	3.3	3.3	2.8	2.3	21-26	77	77
Security Forces (Military Police) (3PXXX)	IRAQ/AFG	LO	4.0	3.8	3.7	3.7	3.2	3.2	3.3	3.1	3.7	3.7	3.2	3.3	2.9	3.8	2.9	2.8	3.0	3.6	10-12	150	3,442
		HI	4.0	3.8	3.7	3.7	3.5	3.6	3.7	3.4	3.9	3.6	3.7	3.5	3.5	3.6	3.3	3.3	3.4	3.5	34-42	150	2,206
	OTHER	LO	4.0	3.8	3.9	3.7	3.8	3.4	3.4	3.3	3.6	3.6	3.3	3.2	3.1	3.2	3.5	3.0	3.1	3.4	11-15	150	2,508
		HI	3.8	3.8	3.8	3.7	3.5	3.6	3.5	3.5	3.5	3.6	3.5	3.4	3.4	3.4	3.2	3.2	3.1	3.1	37-41	150	1,633
Mission Support (3SXXX)	IRAQ/AFG	LO	3.2	2.7	2.9	2.8	2.8	2.9	2.5	2.8	2.9	3.0	2.8	2.7	2.5	2.5	2.7	2.6	2.6	2.6	7-10	92	92
		HI	3.3	3.2	3.3	3.2	3.1	3.2	3.0	3.0	3.1	3.2	2.8	2.9	3.1	2.9	2.9	2.7	2.7	2.5	46-50	150	345
	OTHER	LO	3.9	3.9	3.8	4.0	3.6	3.7	3.3	3.6	3.4	3.8	2.9	3.4	3.1	3.0	3.2	3.0	2.9	2.7	12-16	121	121
		HI	3.5	3.5	3.6	3.4	3.3	3.2	3.1	3.2	3.2	3.2	3.1	3.1	3.1	2.8	3.1	3.1	2.7	2.5	47-54	150	467
Visual Information (3VXXX)	IRAQ/AFG	LO																			3	54	54
		HI	3.5	3.4	3.5	3.4	3.1	3.1	3.2	3.0	3.3	3.3	2.8	3.2	3.0	2.8	3.0	2.7	2.9	2.7	24-30	81	81
	OTHER	LO																			4-6	35	35
		HI	3.6	3.8	3.0	3.5	3.3	3.3	3.3	3.1	3.5	3.3	3.5	3.2	2.8	2.9	3.2	3.3	2.8	2.4	11-16	60	60
Medical (4AXXX)	IRAQ/AFG	LO	3.7	3.6	3.7	3.6	3.4	3.4	3.4	3.3	3.5	3.4	3.3	2.9	3.1	3.0	3.2	3.0	3.2	3.0	16-22	150	157
		HI	3.5	3.4	3.6	3.3	3.2	3.2	3.2	3.0	3.3	3.3	3.2	2.9	3.0	2.8	3.0	2.7	2.9	2.7	44-50	150	189
	OTHER	LO	3.6	3.6	3.7	3.5	3.5	3.4	3.3	3.2	3.2	3.3	3.6	3.0	2.7	2.7	3.2	3.2	2.7	2.4	14-20	150	155
		HI	3.3	3.4	3.5	3.3	3.2	3.2	2.9	3.1	3.0	3.0	2.9	3.0	2.9	2.7	3.0	2.8	2.5	2.4	32-41	150	250
Bioenvironmental Engineering (4BXXX)	IRAQ/AFG	LO																			1-2	8	12
		HI	4.1	3.8	3.7	3.8	3.9	3.8	3.6	3.7	3.6	3.8	3.9	3.5	3.6	3.4	3.6	3.4	3.2	3.0	7	14	25
	OTHER	LO																			0	16	25
		HI																			2-3	19	39
Mental Health Services (4CXXX)	IRAQ/AFG	LO																			1-3	15	21
		HI	3.3	3.0	3.5	3.1	2.8	2.6	2.8	2.6	2.8	3.3	2.9	2.5	2.6	2.6	2.5	2.3	2.4	2.1	5-7	22	40
	OTHER	LO																			2	22	30
		HI																			4-6	12	23
Diet Therapy (4DXXX)	IRAQ/AFG	LO																			0-2	13	22
		HI																			0-1	8	13
	OTHER	LO																			0	5	9
		HI																			0	1	1
Public Health (4EXXX)	IRAQ/AFG	LO																			1	8	14
		HI																			2	10	21
	OTHER	LO																			2-3	17	23
		HI	3.3	3.4	3.6	3.4	3.3	3.2	3.6	3.2	3.3	3.4	3.2	3.3	3.4	3.0	3.1	3.0	3.4	2.5	5-7	17	47
Cardiopulmonary Laboratory (4HXXX)	IRAQ/AFG	LO																			1-2	29	39
		HI																			3-5	8	27
	OTHER	LO																			1-2	13	18
		HI	3.7	3.8	3.9	3.6	3.6	3.3	3.4	3.3	3.4	3.7	3.9	3.3	3.2	2.8	3.3	3.1	2.7	1.9	7-9	10	27

RAND *MG811-T.E.2c*

Table E.2—continued

2-Digit AFSC	Deployed Location	Grade	Gather/Interpret Info	Apply Social Etiquette—SHORT	Respect Cultural Diffs	Apply Social Etiquette	Self-Initiated Learning—SHORT	Change Behavior—SHORT	Establish Credibility	Change Behavior	Verbal/Nonverbal Comm	Manage Stress	Item: Speak Language	Negotiate with Others	Apply Regional Knowledge	Resolve Conflict	Self-Initiated Learning	Foreign Lang Skills	Influence Others	Establish Authority	# Respondents	Invited Sample	Population
Physical Medicine (4JXXX)	IRAQ/AFG	LO																			0	4	4
		HI																			2-3	10	24
	OTHER	LO																			0	0	0
		HI																			0	3	4
Aerospace Physiology (4MXXX)	IRAQ/AFG	LO																			0	2	5
		HI																			2	2	11
	OTHER	LO																			2	16	20
		HI																			4-6	13	25
Medical Service (4NXXX)	IRAQ/AFG	LO	3.9	3.5	3.6	3.6	3.3	3.1	3.5	3.1	3.5	3.3	3.6	3.0	3.2	3.0	3.0	2.6	3.2	2.7	13-16	150	271
		HI	3.5	3.4	3.6	3.4	3.4	3.3	3.3	3.2	3.6	3.4	3.4	3.1	3.0	2.9	3.1	2.9	3.0	2.9	37-46	150	468
	OTHER	LO	3.2	3.3	3.3	3.2	3.0	2.9	2.8	2.9	2.9	3.0	3.3	2.8	2.7	2.8	2.9	2.7	2.5	2.5	13-16	150	274
		HI	3.3	3.3	3.4	3.2	3.3	3.1	3.0	2.9	3.0	3.1	3.2	2.9	2.9	2.7	3.1	3.0	2.6	2.3	45-51	150	532
Pharmacy (4PXXX)	IRAQ/AFG	LO																			2	14	20
		HI																			2-3	14	32
	OTHER	LO																			1	6	9
		HI	4.1	4.0	4.2	3.8	3.8	3.7	3.5	3.5	3.7	3.6	3.9	3.4	3.2	2.9	3.4	3.4	3.1	2.5	6-8	23	36
Diagnostic Imaging (4RXXX)	IRAQ/AFG	LO																			2-3	21	30
		HI	3.2	3.2	3.2	3.0	3.0	3.1	3.1	2.9	3.2	3.0	2.7	2.6	2.8	2.8	2.9	2.6	2.6	2.8	5-6	17	28
	OTHER	LO																			0	9	14
		HI																			2	11	30
Medical Laboratory (4TXXX)	IRAQ/AFG	LO																			2	14	26
		HI	3.2	2.9	3.1	2.9	2.8	2.7	2.9	2.6	2.9	3.2	2.9	2.4	2.6	2.4	2.6	2.6	2.5	2.3	7-9	23	53
	OTHER	LO																			2-3	17	27
		HI																			3-5	14	32
Optometry (4VXXX)	IRAQ/AFG	LO																			0	2	2
		HI																			0	3	5
	OTHER	LO																			0-1	3	8
		HI																			0	3	8
Dental (4YXXX)	IRAQ/AFG	LO																			0-2	20	30
		HI																			3	19	36
	OTHER	LO																			1-2	26	38
		HI																			4-5	24	37
Paralegal (5JXXX)	IRAQ/AFG	LO																			0	1	1
		HI	3.2	3.5	3.5	3.4	3.2	3.2	3.0	3.0	3.0	3.3	2.9	2.7	2.8	2.5	3.0	2.6	2.4	2.3	21-27	67	67
	OTHER	LO																			0-1	6	6
		HI	3.4	3.7	3.3	3.7	3.3	3.4	2.9	3.2	2.9	3.1	2.8	2.6	2.9	2.6	3.1	2.8	2.4	2.3	11-12	44	44
Chaplain Assistant (5RXXX)	IRAQ/AFG	LO																			2-3	24	24
		HI	3.6	3.6	3.7	3.6	3.3	3.4	3.3	3.1	3.1	3.5	3.1	3.0	3.2	2.8	3.1	2.7	2.7	2.8	9-14	48	48
	OTHER	LO																			3-4	21	21
		HI	3.7	4.0	3.8	3.9	3.3	3.7	3.2	3.4	3.3	3.5	3.2	3.4	3.4	2.5	3.1	2.9	2.6	2.2	10-13	33	33
Contracting (6CXXX)	IRAQ/AFG	LO	4.1	4.1	4.2	4.1	4.0	3.7	4.0	3.7	3.9	3.9	4.1	4.2	3.7	3.6	3.6	3.3	3.3	3.3	7-8	51	51
		HI	3.9	4.1	4.1	4.0	3.7	3.7	3.9	3.6	3.9	3.7	3.3	4.2	3.3	3.3	3.5	2.9	3.3	3.0	53-60	150	211
	OTHER	LO	4.1	4.4	4.3	4.3	3.9	3.8	3.9	3.5	3.9	3.4	3.4	4.2	3.6	2.9	3.6	3.0	3.3	3.1	6-8	59	59
		HI	3.6	3.8	3.8	3.8	3.4	3.5	3.6	3.3	3.6	3.2	3.0	4.0	3.1	3.1	3.2	2.9	3.0	2.7	46-51	150	168

RAND *MG811-T.E.2d*

Table E.2—continued

2-Digit AFSC	Deployed Location	Grade	Gather/Interpret Info	Apply Social Etiquette—SHORT	Respect Cultural Diffs	Apply Social Etiquette	Self-Initiated Learning—SHORT	Change Behavior—SHORT	Establish Credibility	Change Behavior	Verbal/Nonverbal Comm	Manage Stress	Item: Speak Language	Negotiate with Others	Apply Regional Knowledge	Resolve Conflict	Self-Initiated Learning	Foreign Lang Skills	Influence Others	Establish Authority	# Respondents	Invited Sample	Population
Financial (6FXXX)	IRAQ/AFG	LO																			3-4	55	55
		HI	3.5	3.6	3.7	3.5	3.3	3.3	3.3	3.2	3.3	3.2	3.1	3.1	3.1	2.8	3.1	2.8	2.8	2.7	19-24	118	118
	OTHER	LO																			4	103	103
		HI	3.7	3.7	3.8	3.7	3.6	3.6	3.4	3.4	3.4	3.3	3.3	3.5	3.1	2.8	3.4	2.9	2.7	2.4	39-46	150	156
Special Investigations (OSI) (7SXXX)	IRAQ/AFG	LO																			0	0	0
		HI	4.4	4.6	4.2	4.5	3.9	4.2	4.3	4.1	4.4	3.9	4.2	4.3	4.2	3.7	3.6	3.8	3.9	3.4	23-25	121	121
	OTHER	LO																			0	0	0
		HI	4.4	4.5	4.2	4.5	4.4	4.3	4.2	4.2	4.3	4.2	3.8	4.3	4.2	3.7	4.0	3.8	4.0	2.8	7-8	81	81
Enlisted Aide (8AXXX)	IRAQ/AFG	LO																			0	0	0
		HI																			0	1	1
	OTHER	LO																			0	0	0
		HI																			0	9	9
Military Training (8BXXX)	IRAQ/AFG	LO																			0	1	1
		HI																			3-4	16	16
	OTHER	LO																			0	0	0
		HI																			3-4	15	15
Family Support Center (8CXXX)	IRAQ/AFG	LO																			0	0	0
		HI																			0	0	0
	OTHER	LO																			0	0	0
		HI																			1-3	6	6
Linguist Debriefer (8DXXX)	IRAQ/AFG	LO																			0	0	0
		HI	4.5	4.7	4.5	4.7	4.3	4.5	4.3	4.2	4.8	4.2	4.4	4.3	4.3	4.0	3.9	4.0	3.7	3.6	6-8	11	11
	OTHER	LO																			0	0	0
		HI																			0	0	0
First Sergeant (8FXXX)	IRAQ/AFG	LO																			0	0	0
		HI	3.2	3.1	3.1	3.0	2.9	2.9	2.9	2.8	3.0	2.9	2.9	2.7	2.8	2.6	2.7	2.7	2.6	2.5	45-51	141	141
	OTHER	LO																			0	0	0
		HI	3.5	3.6	3.6	3.5	3.2	3.4	3.1	3.2	3.0	3.2	2.9	2.9	2.8	2.9	2.9	2.7	2.6	2.2	44-49	150	189
USAF Honor Guard (8GXXX)	IRAQ/AFG	LO																			0	0	0
		HI																			0	1	1
	OTHER	LO																			0	0	0
		HI																			0	0	0
Postal Specialist (8MXXX)	IRAQ/AFG	LO																			0-3	22	22
		HI	3.1	3.0	3.2	3.0	2.9	3.0	2.8	3.0	2.8	3.0	2.8	2.7	2.9	2.8	2.8	2.7	2.7	2.6	9-13	52	52
	OTHER	LO	3.9	3.7	3.8	3.6	3.7	3.3	3.5	3.1	3.3	3.4	2.9	2.9	2.9	2.9	3.4	2.8	2.9	2.6	9-14	89	89
		HI	3.5	3.5	3.4	3.4	3.3	3.4	3.2	3.3	3.3	3.1	3.3	3.1	3.0	3.0	3.2	3.2	2.9	2.6	31-39	100	100
Recruiter (8RXXX)	IRAQ/AFG	LO																			0	3	3
		HI																			0-1	6	6
	OTHER	LO																			0	4	4
		HI																			1-3	15	15
Professional Military Education Instructor (8TXXX)	IRAQ/AFG	LO																			1	0	0
		HI																			0	1	1
	OTHER	LO																			0	0	0
		HI																			0	3	3

RAND *MG811-T.E.2e*

Table E.2—continued

2-Digit AFSC	Deployed Location	Grade	Gather/Interpret Info	Apply Social Etiquette—SHORT	Respect Cultural Diffs	Apply Social Etiquette	Self-Initiated Learning—SHORT	Change Behavior—SHORT	Establish Credibility	Change Behavior	Verbal/Nonverbal Comm	Manage Stress	Item: Speak Language	Negotiate with Others	Apply Regional Knowledge	Resolve Conflict	Self-Initiated Learning	Foreign Lang Skills	Influence Others	Establish Authority	# Respondents	Invited Sample	Population
Awaiting Retraining (9AXXX)	IRAQ/AFG	LO																			1	1	1
		HI																			0	0	0
	OTHER	LO																			0	4	4
		HI																			0	0	0
Dorm Manager (9DXXX)	IRAQ/AFG	LO																			0	0	0
		HI																			0	9	9
	OTHER	LO																			0	1	1
		III																			0	10	10
Senior Enlisted Advisor (9EXXX)	IRAQ/AFG	LO																			0	0	0
		HI																			0	1	1
	OTHER	LO																			0	0	0
		HI																			3	4	4
Group Superintendent (9GXXX)	IRAQ/AFG	LO																			0	0	0
		HI	3.4	3.2	3.3	3.2	3.0	4.0	2.9	3.0	3.0	3.0	3.2	2.7	3.0	2.8	2.8	2.7	2.7	2.5	9-11	25	25
	OTHER	LO																			0	0	0
		HI	3.7	4.0	3.9	3.8	3.4	3.2	3.5	3.7	3.3	3.4	3.1	3.1	3.2	3.1	3.2	2.9	2.8	2.6	9-10	23	23
Interpreter/Translator (9LXXX)	IRAQ/AFG	LO																			0	0	0
		HI																			0	0	0
	OTHER	LO																			0	0	0
		HI																			0	3	3
Technical Applications Specialist (9SXXX)	IRAQ/AFG	LO																			0	0	0
		HI																			1	0	0
	OTHER	LO																			0	5	5
		HI	3.3	3.8	3.5	3.6	3.7	2.0	3.0	3.0	3.4	2.9	3.2	2.6	2.6	2.6	3.4	2.7	2.0	1.6	5	18	18
Basic Enlisted Trainee (9TXXX)	IRAQ/AFG	LO																			0	13	13
		HI																			0	0	0
	OTHER	LO																			1	8	8
		HI																			1	0	0
Unallotted Airman (9UXXX)	IRAQ/AFG	LO	4.3	4.0	4.2	4.0	3.8	3.6	3.9	3.4	4.1	3.8	3.5	3.3	3.4	3.7	3.6	2.9	3.7	3.9	9-13	0	0
		HI	4.3	3.9	3.9	3.7	3.8	3.4	3.8	3.1	4.2	3.9	4.1	3.6	3.2	3.9	3.5	3.2	3.6	4.1	13-16	0	0
	OTHER	LO	3.9	3.9	3.8	3.9	3.2	3.4	3.4	3.3	3.8	3.6	3.6	3.1	3.3	3.2	3.0	2.9	3.2	3.4	11-14	0	0
		HI	3.9	3.8	3.9	3.8	3.6	3.6	3.5	3.4	3.8	3.5	3.3	3.4	3.3	3.1	3.2	3.1	3.2	3.2	7-8	0	0
TCN Escort	IRAQ/AFG	LO	3.7	3.3	3.4	3.2	3.1	3.0	3.2	2.9	3.3	3.3	3.2	2.9	3.0	3.2	2.8	2.8	3.1	3.3	22-27	0	0
		HI	4.1	3.6	3.9	3.6	3.6	3.2	3.7	3.0	3.8	3.5	3.6	3.2	3.1	3.5	3.3	3.1	3.4	3.6	20-23	0	0
	OTHER	LO	3.9	3.9	3.8	3.8	3.4	3.5	3.5	3.4	3.7	3.5	3.4	3.3	3.2	3.3	3.3	2.9	3.2	3.2	35-43	0	0
		HI	3.9	3.9	3.9	3.9	3.6	3.5	3.6	3.5	3.8	3.4	3.3	3.3	3.2	3.2	3.3	2.8	3.3	3.3	22-28	0	0

RAND *MG811-T.E.2f*

Table E.3
Officer Over-Career Training, Proportion Reporting Being Trained, by AFSC, Grade, and Deployment Subgroups

2-Digit AFSC	Deployed Location	Grade	Gather/Interpret Info	Respect Cultural Diffs	Apply Social Etiquette	Establish Credibility	Change Behavior	Verbal/Nonverbal Comm	Manage Stress	Negotiate with Others	Apply Regional Knowledge	Resolve Conflict	Self-Initiated Learning	Foreign Lang Skills	Influence Others	Establish Authority	# Respondents	Invited Sample	Population
All Deployed Personnel	All Deployed Personnel	All Deployed Personnel	.66	.75	.70	.61	.70	.51	.72	.55	.70	.56	.76	.39	.56	.56	5,992-6,018	23,020	88,304
Commander (10XX)	IRAQ/AFG	LO															0	0	0
		HI															4	18	18
	OTHER	LO															0	0	0
		HI	1.00	1.00	.88	.75	.88	.88	1.00	.75	1.00	.75	1.00	.63	.75	.75	8	32	32
Pilot (11XX)	IRAQ/AFG	LO	.59	.63	.72	.50	.66	.44	.72	.47	.75	.42	.78	.34	.44	.44	31-32	150	700
		HI	.67	.71	.67	.58	.66	.48	.69	.48	.71	.56	.73	.47	.55	.52	61-62	150	392
	OTHER	LO	.71	.75	.84	.72	.78	.63	.77	.69	.78	.69	.78	.53	.66	.63	31-32	150	2,711
		HI	.67	.79	.75	.58	.75	.63	.77	.54	.72	.54	.79	.42	.54	.49	57	150	962
Navigator (12XX)	IRAQ/AFG	LO	.70	.73	.70	.63	.79	.55	.76	.61	.85	.58	.85	.55	.59	.61	32-33	150	254
		HI	.53	.58	.45	.37	.50	.37	.55	.32	.63	.37	.66	.21	.37	.34	38	150	161
	OTHER	LO	.66	.72	.66	.62	.66	.38	.62	.55	.76	.59	.76	.34	.59	.48	29	150	900
		HI	.73	.80	.76	.65	.76	.62	.70	.62	.75	.61	.77	.49	.62	.62	71	150	372
Space, Missile, and Command and Control (13XX)	IRAQ/AFG	LO	.61	.71	.70	.48	.68	.42	.71	.52	.81	.55	.71	.39	.52	.52	30-31	150	153
		HI	.59	.77	.68	.66	.73	.45	.86	.50	.82	.50	.82	.32	.50	.50	22	70	70
	OTHER	LO	.62	.69	.69	.55	.62	.45	.59	.41	.62	.45	.72	.45	.45	.45	29	150	413
		HI	.63	.77	.73	.52	.72	.41	.69	.44	.72	.44	.70	.25	.44	.42	64	150	184
Intelligence (14XX)	IRAQ/AFG	LO	.68	.72	.52	.46	.60	.42	.62	.36	.80	.38	.86	.28	.40	.40	50	150	236
		HI	.92	.92	.70	.58	.83	.50	.79	.46	.88	.58	1.00	.29	.50	.46	23-24	68	68
	OTHER	LO	.76	.76	.76	.55	.76	.55	.71	.43	.83	.43	.83	.36	.50	.45	41-42	150	427
		HI	.76	.95	.82	.51	.85	.54	.82	.46	.87	.46	.92	.38	.46	.49	38-39	108	108
Weather (15XX)	IRAQ/AFG	LO	.83	.83	.83	.50	.67	.50	.67	.33	.83	.33	.50	.33	.33	.33	6	37	37
		HI	.57	.57	.71	.57	.71	.57	.86	.57	.71	.57	.71	.57	.57	.57	7	15	15
	OTHER	LO	.73	.82	.55	.64	.82	.36	.73	.64	.82	.64	.91	.20	.64	.64	10-11	38	38
		HI															0-4	9	9
International Affairs/FAO (16XX)	IRAQ/AFG	LO	.54	.62	.54	.38	.54	.31	.54	.38	.62	.38	.77	.31	.38	.38	13	6	6
		HI	.57	.64	.53	.51	.62	.40	.55	.47	.62	.49	.79	.30	.45	.47	47	39	39
	OTHER	LO	.67	.50	.67	.67	.67	.67	.67	.50	.50	.50	.67	.67	.50	.50	6	5	5
		HI	.41	.56	.50	.56	.39	.44	.39	.33	.50	.39	.67	.28	.39	.28	17-18	43	43
Logistics Commander (20XX)	IRAQ/AFG	LO															0	0	0
		HI															4	18	18
	OTHER	LO															0	0	0
		HI	.50	.83	.50	.50	.50	.33	.50	.33	.50	.33	.50	.33	.33	.33	6	10	10

RAND *MG811-T.E.3a*

Table E.3—continued

2-Digit AFSC	Deployed Location	Grade	Gather/Interpret Info	Respect Cultural Diffs	Apply Social Etiquette	Establish Credibility	Change Behavior	Verbal/Nonverbal Comm	Manage Stress	Negotiate with Others	Apply Regional Knowledge	Resolve Conflict	Self-Initiated Learning	Foreign Lang Skills	Influence Others	Establish Authority	# Respondents	Invited Sample	Population
Aircraft Maintenance (21XX)	IRAQ/AFG	LO	.56	.50	.53	.44	.50	.36	.53	.42	.50	.42	.56	.33	.42	.42	**36**	150	295
		HI	.63	.76	.65	.59	.67	.54	.67	.56	.76	.56	.80	.39	.54	.54	**54**	133	133
	OTHER	LO	.64	.66	.61	.47	.61	.42	.67	.42	.69	.42	.67	.36	.43	.44	**35-36**	150	470
		HI	.70	.72	.74	.59	.72	.53	.74	.57	.72	.57	.80	.47	.55	.51	**46-47**	150	153
Support Commander (30XX)	IRAQ/AFG	LO															**0**	0	0
		HI	.70	.78	.80	.80	.80	.60	.80	.70	.80	.80	.90	.20	.80	.80	**9-10**	26	20
	OTHER	LO															**0**	0	0
		HI	.78	.89	1.00	.89	.89	.67	1.00	.78	1.00	.78	1.00	.67	.78	.78	**9**	26	26
Security Forces (31XX)	IRAQ/AFG	LO	.77	.74	.65	.65	.68	.52	.68	.61	.77	.61	.71	.45	.61	.65	**31**	150	178
		HI	.83	1.00	.77	.67	.85	.54	.77	.67	.83	.67	.83	.23	.75	.75	**12-13**	45	45
	OTHER	LO	1.00	.89	1.00	1.00	.89	.67	.78	.67	1.00	.89	1.00	.33	.78	.78	**9**	50	50
		HI															**2**	12	12
Civil Engineer (32XX)	IRAQ/AFG	LO	.56	.73	.70	.62	.68	.53	.71	.53	.74	.53	.74	.38	.53	.56	**33-34**	150	218
		HI	.71	.90	.83	.83	.81	.69	.81	.81	.86	.76	.86	.43	.79	.73	**41-42**	98	98
	OTHER	LO	.75	.79	.71	.67	.67	.42	.67	.54	.67	.54	.92	.33	.58	.54	**24**	139	139
		HI	.64	.80	.93	.60	.87	.67	.60	.53	.53	.53	.80	.50	.47	.53	**14-15**	48	48
Comm-Information Systems (33XX)	IRAQ/AFG	LO	.53	.67	.63	.53	.63	.44	.57	.51	.60	.51	.71	.37	.53	.56	**42-43**	150	317
		HI	.61	.68	.59	.57	.66	.45	.70	.48	.70	.55	.72	.36	.55	.48	**43-44**	91	91
	OTHER	LO	.43	.57	.64	.36	.54	.25	.61	.29	.54	.36	.61	.21	.29	.32	**28**	150	397
		HI	.67	.73	.73	.54	.69	.49	.73	.48	.62	.50	.73	.35	.50	.48	**51-52**	94	94
Services (34XX)	IRAQ/AFG	LO	.55	.65	.55	.40	.60	.40	.50	.30	.55	.35	.70	.30	.35	.35	**20**	45	45
		HI															**3**	15	15
	OTHER	LO	.68	.68	.63	.68	.63	.47	.74	.61	.63	.47	.74	.32	.58	.58	**18-19**	87	87
		HI	1.00	1.00	1.00	.86	1.00	.86	.86	.86	.86	.86	.86	.71	.86	.86	**7**	18	18
Public Affairs (35XX)	IRAQ/AFG	LO	1.00	1.00	1.00	.83	1.00	.83	1.00	.67	1.00	.83	1.00	.67	.83	.83	**6**	32	32
		HI	.71	.86	.57	.57	.57	.71	.71	.57	.57	.57	.71	.14	.57	.57	**7**	11	11
	OTHER	LO	.70	.80	.80	.70	.80	.80	.80	.70	.80	.70	.90	.60	.70	.80	**10**	40	40
		HI															**4-5**	25	25
Personnel (36XX)	IRAQ/AFG	LO															**0**	26	26
		HI															**0**	15	15
	OTHER	LO															**0**	41	41
		HI															**0**	9	9
Manpower-Personnel (37XX)	IRAQ/AFG	LO	.56	.78	.56	.39	.67	.33	.72	.33	.61	.33	.72	.33	.33	.33	**18**	51	51
		HI	.52	.61	.48	.48	.48	.35	.57	.35	.48	.39	.61	.13	.35	.35	**23**	36	36
	OTHER	LO	.74	.76	.67	.67	.76	.50	.81	.60	.74	.64	.79	.38	.62	.62	**42**	83	83
		HI	.50	.64	.64	.50	.57	.50	.64	.50	.54	.50	.79	.36	.43	.43	**13-14**	31	31

RAND *MG811-T.E.3b*

Table E.3—continued

2-Digit AFSC	Deployed Location	Grade	Gather/Interpret Info	Respect Cultural Diffs	Apply Social Etiquette	Establish Credibility	Change Behavior	Verbal/Nonverbal Comm	Manage Stress	Negotiate with Others	Apply Regional Knowledge	Resolve Conflict	Self-Initiated Learning	Foreign Lang Skills	Influence Others	Establish Authority	# Respondents	Invited Sample	Population
Manpower (38XX)	IRAQ/AFG	LO															0	3	3
		HI															0	1	1
	OTHER	LO															0	13	13
		HI															0	3	3
Medical Commander (40XX)	IRAQ/AFG	LO															0	0	0
		HI	1.00	1.00	.60	.80	1.00	.40	1.00	.80	1.00	.80	1.00	.40	.80	.80	5	7	7
	OTHER	LO															0	0	0
		HI	.57	.71	.71	.57	.57	.57	.57	.57	.43	.43	.71	.29	.57	.43	7	14	14
Health Services Administrator (41XX)	IRAQ/AFG	LO	.22	.33	.33	.33	.22	.22	.44	.22	.33	.22	.44	.22	.22	.22	9	40	40
		HI	.75	.88	.81	.81	.81	.63	.88	.75	.81	.81	.81	.63	.75	.75	16	44	44
	OTHER	LO	.63	.75	.50	.63	.50	.38	.75	.50	.50	.50	.75	.38	.38	.38	8	44	44
		HI	.42	.75	.50	.42	.42	.25	.67	.33	.67	.42	.58	.25	.33	.33	12	37	37
Biomedical Clinician (42XX)	IRAQ/AFG	LO	.71	.86	.81	.71	.71	.57	.90	.70	.67	.71	.81	.57	.71	.67	20-21	70	70
		HI	.78	.56	.56	.56	.44	.44	.78	.44	.56	.44	.89	.33	.44	.56	9	34	34
	OTHER	LO															3	48	48
		HI															4	17	17
Biomedical Specialists (43XX)	IRAQ/AFG	LO	.38	.38	.38	.38	.38	.38	.38	.38	.38	.30	.50	.38	.38	.38	8	32	32
		HI	.57	.52	.57	.43	.52	.43	.52	.38	.62	.38	.71	.29	.48	.43	21	49	49
	OTHER	LO	.69	.77	.85	.46	.69	.46	.77	.46	.54	.46	.77	.23	.46	.46	13	34	34
		HI	.67	.73	.64	.59	.64	.41	.77	.50	.73	.45	.77	.18	.41	.41	21-22	42	42
Physician (44XX)	IRAQ/AFG	LO															3-4	46	46
		HI	.55	.66	.59	.47	.56	.38	.63	.47	.68	.47	.69	.34	.47	.53	31-32	120	120
	OTHER	LO															0-3	48	48
		HI	.71	.81	.67	.57	.71	.62	.76	.52	.67	.57	.71	.38	.52	.52	21	77	77
Surgery (45XX)	IRAQ/AFG	LO															1	16	16
		HI	.78	.76	.67	.67	.72	.39	.67	.50	.72	.50	.83	.33	.56	.56	17-18	113	113
	OTHER	LO															0	4	4
		HI	.75	.88	.75	.88	.88	.63	.88	.63	.88	.88	.88	.00	.00	.76	8	38	38
Nurse (46XX)	IRAQ/AFG	LO	.43	.63	.54	.45	.50	.31	.54	.35	.46	.39	.67	.22	.37	.39	48-50	150	258
		HI	.66	.74	.77	.55	.70	.55	.77	.51	.74	.49	.79	.49	.57	.51	46-47	99	99
	OTHER	LO	.65	.70	.67	.59	.72	.59	.67	.54	.70	.57	.70	.50	.59	.59	46	150	269
		HI	.67	.88	.82	.64	.76	.58	.79	.61	.76	.58	.91	.52	.61	.67	33	94	94
Dental (47XX)	IRAQ/AFG	LO															0	4	4
		HI															0-4	12	12
	OTHER	LO															0-1	8	8
		HI	.55	.73	.55	.55	.64	.45	.55	.45	.64	.45	.73	.36	.45	.55	11	24	24

RAND *MG811-T.E.3c*

Table E.3—continued

2-Digit AFSC	Deployed Location	Grade	Gather/Interpret Info	Respect Cultural Diffs	Apply Social Etiquette	Establish Credibility	Change Behavior	Verbal/Nonverbal Comm	Manage Stress	Negotiate with Others	Apply Regional Knowledge	Resolve Conflict	Self-Initiated Learning	Foreign Lang Skills	Influence Others	Establish Authority	# Respondents	Invited Sample	Population
Aerospace Medicine (48XX)	IRAQ/AFG	LO															**0-4**	26	26
		HI	.65	.76	.82	.59	.82	.53	.65	.59	.71	.53	.76	.35	.59	.59	**17**	51	51
	OTHER	LO	.50	.67	.67	.50	.67	.33	.50	.50	.67	.33	.83	.33	.33	.33	**6**	57	57
		HI	.68	.74	.77	.57	.74	.65	.74	.65	.68	.55	.73	.45	.58	.58	**30-31**	81	81
Law (51XX)	IRAQ/AFG	LO	.75	1.00	.88	.75	.88	.63	.88	.63	.88	.75	.88	.38	.75	.75	**8**	56	56
		HI	.67	.76	.62	.60	.71	.62	.62	.57	.67	.52	.71	.48	.62	.57	**20-21**	43	43
	OTHER	LO															**4**	37	37
		HI	.69	.81	.81	.69	.81	.56	.81	.69	.80	.63	.88	.63	.63	.63	**15-16**	49	49
Chaplain (52XX)	IRAQ/AFG	LO	.67	.73	.67	.60	.80	.40	.73	.47	.80	.47	.80	.40	.53	.47	**15**	43	43
		HI	.50	.78	.78	.61	.61	.39	.78	.44	.61	.56	.94	.22	.50	.47	**17-18**	49	49
	OTHER	LO	.73	.82	.64	.55	.82	.45	.73	.45	.73	.36	.73	.36	.45	.27	**11**	43	43
		HI	.70	.80	.90	.40	.90	.33	.70	.30	.70	.40	.90	.30	.40	.40	**9-10**	25	25
Program Director (60XX)	IRAQ/AFG	LO															**0**	0	0
		HI															**0**	1	1
	OTHER	LO															**0**	0	0
		HI															**0**	0	0
Scientific/Research (61XX)	IRAQ/AFG	LO															**0-3**	9	9
		HI															**3-4**	4	4
	OTHER	LO	.83	.83	.83	.83	.83	.83	.83	.83	.83	.83	.83	.83	.83	.83	**6**	15	15
		HI	.43	.57	.57	.29	.43	.29	.43	.29	.43	.29	.43	.29	.29	.29	**7**	16	16
Developmental Engineering (62XX)	IRAQ/AFG	LO	.67	.67	.33	.67	.67	.17	.67	.50	.67	.50	.67	.17	.50	.33	**6**	21	21
		HI															**0-1**	10	10
	OTHER	LO	.86	.71	.86	.57	.86	.57	.57	.57	.86	.57	.86	.50	.57	.57	**6-7**	31	31
		HI															**0**	5	5
Acquisition (63XX)	IRAQ/AFG	LO	.88	1.00	.88	.63	.88	.63	.75	.50	1.00	.63	.75	.50	.63	.63	**8**	33	33
		HI	.67	.63	.67	.56	.67	.56	.63	.44	.67	.44	.67	.44	.44	.44	**8-9**	42	42
	OTHER	LO															**2**	31	31
		HI															**3**	17	17
Contracting (64XX)	IRAQ/AFG	LO	.63	.78	.72	.66	.66	.59	.72	.66	.75	.56	.84	.25	.59	.53	**32**	111	111
		HI	.71	.86	.86	.86	.86	.62	.76	.86	.76	.76	.86	.33	.76	.57	**21**	52	52
	OTHER	LO	.71	.86	.82	.82	.82	.55	.91	.82	.73	.68	.82	.32	.73	.64	**21-22**	69	69
		HI															**3**	4	4
Finance (65XX)	IRAQ/AFG	LO	.75	.75	.38	.63	.75	.38	.88	.50	.63	.63	.75	.25	.50	.63	**8**	43	43
		HI	.60	.60	.40	.60	.60	.20	.60	.40	.60	.40	.60	.20	.40	.40	**5**	37	37
	OTHER	LO	.60	.80	.60	.60	.60	.40	.40	.20	.60	.20	.80	.20	.20	.40	**5**	22	22
		HI	.88	.88	.88	.88	.88	.75	.88	.88	.75	.88	.88	.50	.88	.88	**8**	17	17

RAND *MG811-T.E.3d*

Table E.3—continued

2-Digit AFSC	Deployed Location	Grade	Gather/Interpret Info	Respect Cultural Diffs	Apply Social Etiquette	Establish Credibility	Change Behavior	Verbal/Nonverbal Comm	Manage Stress	Negotiate with Others	Apply Regional Knowledge	Resolve Conflict	Self-Initiated Learning	Foreign Lang Skills	Influence Others	Establish Authority	# Respondents	Invited Sample	Population
Special Investigations (71XX)	IRAQ/AFG	LO															4	30	30
		HI															4	14	14
	OTHER	LO															4	17	17
		HI															0-1	4	4
Commander—USAFA Cadet Squadron (80XX)	IRAQ/AFG	LO															0	0	0
		HI															0	0	0
	OTHER	LO															0	0	0
		HI															0	1	1
Instructor (81XX)	IRAQ/AFG	LO															1	4	4
		HI															3	8	8
	OTHER	LO															0	6	6
		HI															0	8	8
Academic Program Manager (82XX)	IRAQ/AFG	LO															0	1	1
		HI															0	3	3
	OTHER	LO															0	1	1
		HI															0	3	3
Recruiting Service (83XX)	IRAQ/AFG	LO															0	4	4
		HI															0	0	0
	OTHER	LO															0	2	2
		HI															0	0	0
Historian (84XX)	IRAQ/AFG	LO															0	0	0
		HI															0	0	0
	OTHER	LO															0	0	0
		HI															0	0	0
USAF Honor Guard (85XX)	IRAQ/AFG	LO															0	0	0
		HI															0	0	0
	OTHER	LO															0	0	0
		HI															0	1	1
Command and Control (86XX)	IRAQ/AFG	LO															0	2	2
		HI															0-3	3	3
	OTHER	LO															3	3	3
		HI															0-1	5	5
Inspector General (87XX)	IRAQ/AFG	LO															0	0	0
		HI															0	0	0
	OTHER	LO															0	0	0
		HI															0	1	1

RAND *MG811-T.E.3e*

Table E.3—continued

2-Digit AFSC	Deployed Location	Grade	Gather/Interpret Info	Respect Cultural Diffs	Apply Social Etiquette	Establish Credibility	Change Behavior	Verbal/Nonverbal Comm	Manage Stress	Negotiate with Others	Apply Regional Knowledge	Resolve Conflict	Self-Initiated Learning	Foreign Lang Skills	Influence Others	Establish Authority	# Respondents	Invited Sample	Population
Protocol/Aide (88XX)	IRAQ/AFG	LO															0-2	1	1
		HI															0-1	0	0
	OTHER	LO															3	0	0
		HI															0	0	0
General Officer (90XX)	IRAQ/AFG	LO															0	0	0
		HI															3	10	10
	OTHER	LO															0	0	0
		HI															0	11	11
Commander (91XX)	IRAQ/AFG	LO															0	0	0
		HI															4	13	13
	OTHER	LO															0	0	0
		HI	.78	.78	.78	.78	.78	.67	.78	.78	.67	.78	.78	.44	.78	.78	9	18	18
Student (92XX)	IRAQ/AFG	LO															0	22	22
		HI															0	7	7
	OTHER	LO															0-2	65	65
		HI															0	12	12
Patient (93XX)	IRAQ/AFG	LO															0	0	0
		HI															0	0	0
	OTHER	LO															0	0	0
		HI															0	0	0
Nuclear Weapons Custodian (94XX)	IRAQ/AFG	LO															0	0	0
		HI															0	0	0
	OTHER	LO															0	0	0
		HI															0	0	0
Non-extended Active Duty (95XX)	IRAQ/AFG	LO															0	0	0
		HI															0	0	0
	OTHER	LO															0	0	0
		HI															0	1	1
Unclassified Officer (96XX)	IRAQ/AFG	LO															0	0	0
		HI															0	0	0
	OTHER	LO															1	2	2
		HI															0	0	0
Executive Officer above Wing Level (97XX)	IRAQ/AFG	LO															1	3	3
		HI															2	12	12
	OTHER	LO															0-2	6	6
		HI															0	7	7

RAND *MG811-T.E.3f*

Table E.4
Enlisted Over-Career Training, Proportion Reporting Being Trained, by AFSC, Grade, and Deployment Subgroups

2-Digit AFSC	Deployed location	Grade	Gather/Interpret Info	Respect Cultural Diffs	Apply Social Etiquette	Establish Credibility	Change Behavior	Verbal/Nonverbal Comm	Manage Stress	Negotiate with Others	Apply Regional Knowledge	Resolve Conflict	Self-Initiated Learning	Foreign Lang Skills	Influence Others	Establish Authority	# Respondents	Invited Sample	Population
All Deployed Personnel	All Deployed Personnel	All Deployed Personnel	.66	.75	.70	.61	.70	.51	.72	.55	.70	.56	.76	.39	.56	.56	5,992-6,018	23,020	88,304
Aircrew Operations (1AXXX)	IRAQ/AFG	LO	.67	.73	.73	.73	.73	.53	.73	.73	.80	.67	.80	.47	.73	.73	15	150	218
		HI	.42	.50	.58	.37	.53	.32	.53	.32	.55	.32	.63	.26	.34	.34	38-39	150	633
	OTHER	LO	.79	.88	.83	.71	.83	.54	.79	.71	.83	.67	.83	.42	.70	.58	23-24	150	1,248
		HI	.74	.79	.76	.64	.81	.55	.71	.57	.79	.57	.79	.40	.60	.61	41-42	150	1,431
Command & Control Systems Operations (1CXXX)	IRAQ/AFG	LO	.63	.79	.53	.74	.72	.53	.61	.58	.63	.68	.63	.47	.63	.63	18-19	150	859
		HI	.66	.74	.71	.64	.76	.58	.76	.59	.72	.62	.80	.54	.62	.62	49-50	150	983
	OTHER	LO	.80	.90	.90	.80	.85	.70	.85	.80	.90	.75	.90	.55	.85	.75	20	150	554
		HI	.69	.81	.74	.60	.75	.55	.79	.50	.76	.55	.77	.44	.56	.53	61-62	150	676
Intelligence (1NXXX)	IRAQ/AFG	LO	.60	.67	.73	.53	.60	.60	.60	.53	.73	.53	.93	.40	.60	.60	15	150	290
		HI	.67	.75	.61	.46	.65	.37	.68	.35	.72	.37	.82	.25	.39	.36	56-57	150	313
	OTHER	LO	.65	.75	.70	.50	.70	.45	.75	.40	.80	.40	.80	.25	.45	.40	20	150	436
		HI	.67	.75	.71	.54	.71	.48	.75	.52	.79	.52	.77	.42	.55	.52	47-48	150	437
Safety (1SXXX)	IRAQ/AFG	LO															0	0	0
		HI	.53	.67	.47	.47	.47	.20	.53	.33	.53	.40	.73	.13	.40	.40	15	39	39
	OTHER	LO															0	1	1
		HI	.57	.64	.71	.64	.57	.21	.64	.29	.64	.36	.64	.14	.36	.43	14	41	41
Aircrew Protection (1TXXX)	IRAQ/AFG	LO	.67	.67	.67	.56	.56	.56	.67	.56	.56	.56	.67	.56	.56	.56	9	115	115
		HI	.52	.63	.67	.44	.59	.37	.56	.41	.63	.44	.63	.33	.44	.44	27	118	118
	OTHER	LO	.67	.67	.67	.67	.75	.50	.58	.67	.67	.58	.67	.33	.67	.67	12	150	276
		HI	.64	.67	.73	.53	.67	.49	.71	.47	.64	.49	.78	.38	.44	.44	45	150	290
Weather (1WXXX)	IRAQ/AFG	LO	.50	.67	.67	.50	.50	.50	.50	.50	.67	.50	.67	.50	.50	.50	6	68	68
		HI	.59	.69	.61	.51	.59	.42	.63	.44	.67	.44	.75	.31	.49	.49	58-59	150	217
	OTHER	LO															4	54	54
		HI	.64	.84	.75	.50	.68	.48	.75	.45	.82	.49	.82	.36	.50	.45	43-44	127	127
Manned Aerospace Maintenance (2AXXX)	IRAQ/AFG	LO	.50	.60	.70	.60	.60	.30	.60	.50	.50	.60	.50	.20	.50	.50	10	150	2,505
		HI	.54	.72	.59	.48	.62	.44	.88	.45	.60	.18	.77	.34	.50	.55	59-62	150	2,396
	OTHER	LO	.73	.83	.75	.67	.67	.50	.75	.75	.75	.67	.92	.33	.67	.67	11-12	150	5,727
		HI	.53	.68	.68	.60	.58	.54	.68	.58	.67	.54	.68	.47	.54	.54	57	150	5,923
Communications & Electronics (2EXXX)	IRAQ/AFG	LO	.65	.83	.74	.65	.70	.52	.83	.57	.74	.64	.74	.48	.70	.61	22-23	150	705
		HI	.56	.73	.69	.57	.67	.51	.74	.51	.65	.52	.73	.42	.51	.51	77-78	150	596
	OTHER	LO	.57	.73	.69	.53	.63	.30	.73	.47	.57	.50	.63	.17	.47	.47	29-30	150	654
		HI	.63	.72	.74	.63	.75	.47	.75	.56	.67	.58	.75	.39	.53	.54	57	150	689
Fuels (2FXXX)	IRAQ/AFG	LO	.68	.68	.64	.60	.68	.40	.76	.56	.68	.60	.76	.36	.64	.60	25	150	209
		HI	.63	.70	.70	.58	.64	.48	.68	.57	.65	.53	.77	.40	.55	.57	59-60	150	264
	OTHER	LO	.67	.80	.80	.67	.67	.73	.67	.73	.73	.67	.73	.53	.73	.73	15	150	668
		HI	.74	.81	.76	.68	.74	.57	.83	.57	.77	.61	.77	.41	.57	.58	67-69	150	709

RAND *MG811-T.E.4a*

Table E.4—continued

2-Digit AFSC	Deployed location	Grade	Gather/Interpret Info	Respect Cultural Diffs	Apply Social Etiquette	Establish Credibility	Change Behavior	Verbal/Nonverbal Comm	Manage Stress	Negotiate with Others	Apply Regional Knowledge	Resolve Conflict	Self-Initiated Learning	Foreign Lang Skills	Influence Others	Establish Authority	# Respondents	Invited Sample	Population
Logistics Plans (2GXXX)	IRAQ/AFG	LO															0-2	22	22
		HI	.79	.79	.83	.72	.76	.62	.83	.69	.79	.69	.83	.55	.69	.69	29	87	87
	OTHER	LO	.71	.71	.71	.57	.71	.57	.71	.57	.71	.71	.57	.43	.57	.57	7	37	37
		HI	.69	.79	.77	.62	.69	.51	.77	.49	.76	.46	.79	.33	.50	.51	38-39	119	119
Missile & Space Systems Maintenance (2MXXX)	IRAQ/AFG	LO															2-3	59	59
		HI															4	22	22
	OTHER	LO															0	39	39
		HI															1	7	7
Precision Measurement Equipment Lab (2PXXX)	IRAQ/AFG	LO															0	5	5
		HI															0-2	5	5
	OTHER	LO															0-5	42	42
		HI	.47	.67	.60	.37	.63	.33	.53	.33	.53	.37	.47	.30	.33	.33	30	61	61
Maintenance Management Systems (2RXXX)	IRAQ/AFG	LO															0-3	42	42
		HI	.60	.80	.60	.60	.70	.30	.60	.50	.70	.60	.70	.20	.70	.70	10	35	35
	OTHER	LO	.83	1.00	.83	.83	1.00	.50	1.00	.50	.83	.67	.83	.33	.50	.67	6	92	92
		HI	.67	.74	.89	.56	.70	.52	.85	.41	.70	.48	.77	.37	.44	.59	26-27	100	100
Supply (2SXXX)	IRAQ/AFG	LO	1.00	1.00	1.00	.82	1.00	.82	1.00	.91	1.00	.82	1.00	.55	.82	.91	11	150	374
		HI	.59	.67	.53	.53	.63	.53	.63	.47	.57	.47	.57	.37	.47	.43	48-49	150	643
	OTHER	LO	.77	.77	.62	.69	.69	.54	.77	.62	.77	.69	.77	.46	.69	.69	13	150	650
		HI	.66	.68	.79	.68	.74	.55	.74	.61	.71	.63	.74	.42	.62	.61	38-39	150	1,067
Transportation & Vehicle Maintenance (2TXXX)	IRAQ/AFG	LO	.75	.79	.71	.79	.71	.58	.79	.70	.71	.79	.79	.54	.79	.79	23-24	150	1,232
		HI	.70	.72	.71	.65	.71	.54	.73	.60	.76	.66	.80	.43	.64	.66	65-67	150	1,219
	OTHER	LO	.62	.69	.62	.62	.69	.62	.69	.62	.62	.62	.62	.50	.62	.62	12-13	150	1,207
		HI	.80	.84	.82	.71	.82	.60	.80	.67	.76	.67	.82	.51	.69	.73	44-45	150	1,275
Munitions & Weapons (2WXXX)	IRAQ/AFG	LO	.65	.78	.74	.65	.74	.61	.78	.65	.65	.65	.70	.52	.65	.65	23	150	768
		HI	.61	.71	.64	.53	.67	.45	.80	.51	.63	.53	.65	.37	.57	.53	50-51	150	714
	OTHER	LO	.63	.79	.68	.74	.74	.53	.74	.58	.68	.68	.68	.53	.58	.63	19	150	980
		HI	.55	.73	.73	.50	.63	.50	.63	.50	.57	.50	.60	.44	.50	.52	54-57	150	905
Information Management (3AXXX)	IRAQ/AFG	LO	.74	.74	.74	.68	.68	.53	.74	.68	.74	.68	.74	.47	.74	.74	19	150	220
		HI	.54	.62	.58	.48	.62	.39	.60	.48	.59	.50	.67	.31	.53	.49	57-59	150	539
	OTHER	LO	.72	.72	.68	.71	.72	.58	.74	.72	.72	.71	.83	.53	.72	.72	17-19	150	329
		HI	.48	.67	.57	.43	.65	.42	.59	.42	.54	.35	.70	.28	.38	.35	53-54	150	756
Communications & Computer Systems (3CXXX)	IRAQ/AFG	LO	.59	.59	.73	.57	.68	.50	.82	.64	.64	.64	.59	.50	.57	.68	21-22	150	405
		HI	.57	.67	.57	.45	.65	.37	.63	.37	.57	.48	.63	.26	.43	.43	44-46	150	627
	OTHER	LO	.67	.69	.77	.69	.69	.54	.69	.54	.67	.62	.85	.54	.54	.62	12-13	150	617
		HI	.59	.73	.67	.55	.69	.50	.67	.55	.64	.48	.77	.40	.52	.52	44-45	150	1,026
Civil Engineering (3EXXX)	IRAQ/AFG	LO	.67	.67	.72	.61	.67	.50	.67	.56	.67	.67	.67	.44	.61	.61	18	150	1,839
		HI	.70	.80	.77	.70	.73	.64	.78	.67	.73	.66	.81	.52	.69	.67	62-64	150	1,826
	OTHER	LO	.75	.81	.75	.75	.73	.56	.75	.75	.75	.75	.75	.31	.75	.75	15-16	150	2,039
		HI	.73	.78	.69	.68	.78	.49	.73	.60	.78	.59	.88	.41	.65	.63	39-41	150	1,919

RAND *MG811-T.E.4b*

Table E.4—continued

2-Digit AFSC	Deployed location	Grade	Gather/Interpret Info	Respect Cultural Diffs	Apply Social Etiquette	Establish Credibility	Change Behavior	Verbal/Nonverbal Comm	Manage Stress	Negotiate with Others	Apply Regional Knowledge	Resolve Conflict	Self-Initiated Learning	Foreign Lang Skills	Influence Others	Establish Authority	# Respondents	Invited Sample	Population
Historian (3HXXX)	IRAQ/AFG	LO															0	0	0
		HI															0-1	5	5
	OTHER	LO															0	0	0
		HI															3	7	7
Services (3MXXX)	IRAQ/AFG	LO	.73	.82	.73	.73	.77	.40	.86	.59	.71	.55	.73	.32	.64	.68	20-22	150	208
		HI	.64	.77	.73	.61	.75	.55	.75	.61	.68	.64	.73	.45	.59	.66	44	150	268
	OTHER	LO	.73	.73	.80	.73	.87	.60	.80	.60	.80	.67	.87	.27	.60	.53	15	150	703
		HI	.76	.90	.87	.81	.83	.65	.87	.77	.83	.77	.88	.51	.77	.75	51-52	150	762
Public Affairs (3NXXX)	IRAQ/AFG	LO															3-4	33	33
		HI															32	78	78
	OTHER	LO															3	20	20
		HI	.58	.67	.63	.54	.71	.38	.70	.42	.63	.33	.58	.21	.29	.38	23-24	77	77
Security Forces (Military Police) (3PXXX)	IRAQ/AFG	LO	.92	1.00	.92	.92	.92	.83	.92	.92	.83	1.00	1.00	.75	1.00	1.00	12	150	3,442
		HI	.83	.78	.74	.76	.76	.56	.78	.66	.78	.73	.80	.37	.76	.76	39-41	150	2,206
	OTHER	LO	.88	1.00	.88	.81	.88	.69	.81	.81	.88	.81	.88	.50	.81	.81	16	150	2,508
		HI	.76	.88	.76	.73	.83	.59	.90	.73	.78	.71	.80	.49	.71	.80	40-41	150	1,633
Mission Support (3SXXX)	IRAQ/AFG	LO	.67	.67	.56	.67	.07	.44	.78	.56	.75	.56	.67	.44	.56	.56	8-9	92	92
		HI	.50	.66	.58	.50	.60	.44	.72	.44	.48	.44	.62	.30	.48	.48	50	150	345
	OTHER	LO	.79	.86	.79	.79	.93	.79	.93	.79	.79	.79	.85	.57	.79	.79	13-14	121	121
		HI	.63	.78	.64	.69	.65	.53	.76	.57	.73	.55	.78	.43	.59	.55	50-51	150	467
Visual Information (3VXXX)	IRAQ/AFG	LO															4	54	54
		HI	.71	.82	.64	.64	.75	.63	.79	.64	.68	.57	.79	.39	.61	.61	27-28	81	81
	OTHER	LO															0-5	35	35
		HI	.73	.80	.67	.73	.80	.44	.73	.53	.60	.53	.86	.44	.53	.53	[illegible]	[illegible]	[illegible]
Medical (4AXXX)	IRAQ/AFG	LO	.67	.76	.62	.62	.62	.57	.76	.57	.71	.57	.75	.38	.67	.62	20-21	150	157
		HI	.53	.66	.60	.49	.55	.32	.66	.42	.60	.40	.75	.23	.42	.46	52-53	150	189
	OTHER	LO	.70	.75	.65	.60	.60	.50	.75	.60	.70	.55	.75	.35	.65	.60	20	150	155
		HI	.73	.70	.00	.57	.68	.49	.68	.51	.68	.43	.68	.35	.46	.46	37	150	250
Bioenvironmental Engineering (4BXXX)	IRAQ/AFG	LO															2	8	12
		HI	.71	.71	.57	.71	.71	.43	.86	.57	.71	.57	.86	.29	.57	.57	7	14	25
	OTHER	LO															0	16	25
		HI															3	19	39
Mental Health Services (4CXXX)	IRAQ/AFG	LO															0-3	15	21
		HI	.57	.71	.57	.57	.57	.43	.71	.43	.57	.43	.71	.29	.57	.43	7	22	40
	OTHER	LO															2	22	30
		HI	.33	.83	.50	.50	.33	.33	.67	.33	.50	.33	.67	.17	.50	.33	6	12	23
Diet Therapy (4DXXX)	IRAQ/AFG	LO															0	13	22
		HI															1	8	13
	OTHER	LO															0	5	9
		HI															0	1	1

RAND *MG811-T.E.4c*

Table E.4—continued

2-Digit AFSC	Deployed location	Grade	Gather/Interpret Info	Respect Cultural Diffs	Apply Social Etiquette	Establish Credibility	Change Behavior	Verbal/Nonverbal Comm	Manage Stress	Negotiate with Others	Apply Regional Knowledge	Resolve Conflict	Self-Initiated Learning	Foreign Lang Skills	Influence Others	Establish Authority	# Respondents	Invited Sample	Population
Public Health (4EXXX)	IRAQ/AFG	LO															0	8	14
		HI															2	10	21
	OTHER	LO															3	17	23
		HI	.67	.83	.50	.67	.67	.50	.83	.50	.67	.33	.67	.33	.50	.50	6	17	47
Cardiopulmonary Laboratory (4HXXX)	IRAQ/AFG	LO															3	29	39
		HI															4	8	27
	OTHER	LO															2	13	18
		HI	.75	.75	.75	.63	.75	.75	.75	.50	.63	.50	.75	.50	.50	.50	8	10	27
Physical Medicine (4JXXX)	IRAQ/AFG	LO															0	4	4
		HI															2	10	24
	OTHER	LO															0	0	0
		HI															0	3	4
Aerospace Physiology (4MXXX)	IRAQ/AFG	LO															0	2	5
		HI															2	2	11
	OTHER	LO															3	16	20
		HI	.83	1.00	.83	.83	.67	.67	.67	.67	.83	.67	.67	.67	.67	.83	5-6	13	25
Medical Service (4NXXX)	IRAQ/AFG	LO	.73	.93	.87	.80	.87	.64	.87	.73	.87	.80	.80	.53	.73	.73	14-15	150	271
		HI	.68	.78	.66	.57	.74	.49	.77	.55	.70	.55	.77	.38	.57	.60	46-47	150	468
	OTHER	LO	.65	.71	.67	.59	.78	.59	.78	.50	.71	.53	.76	.44	.53	.53	16-18	150	274
		HI	.58	.60	.57	.53	.60	.38	.60	.45	.66	.51	.74	.32	.47	.51	53	150	532
Pharmacy (4PXXX)	IRAQ/AFG	LO															2	14	20
		HI															2	14	32
	OTHER	LO															1	6	9
		HI	.71	.75	.75	.75	.75	.63	.75	.75	.75	.75	.75	.63	.75	.75	7-8	23	36
Diagnostic Imaging (4RXXX)	IRAQ/AFG	LO															3	21	30
		HI	1.00	1.00	1.00	1.00	1.00	1.00	1.00	1.00	1.00	1.00	1.00	.83	1.00	1.00	6	17	28
	OTHER	LO															0	9	14
		HI															0-2	11	30
Medical Laboratory (4TXXX)	IRAQ/AFG	LO															2	14	26
		HI	.56	.78	.67	.56	.67	.44	.78	.44	.67	.44	.78	.33	.44	.44	9	23	53
	OTHER	LO															3	17	27
		HI	.50	.67	.50	.33	.33	.50	.50	.33	.50	.33	.67	.33	.33	.33	6	14	32
Optometry (4VXXX)	IRAQ/AFG	LO															0	2	2
		HI															0	3	5
	OTHER	LO															0-1	3	8
		HI															0	3	8
Dental (4YXXX)	IRAQ/AFG	LO															2	20	30
		HI															2	19	36
	OTHER	LO															2	26	38
		HI	1.00	1.00	1.00	1.00	1.00	.80	1.00	1.00	1.00	1.00	1.00	.60	1.00	1.00	5	24	37

RAND *MG811-T.E.4d*

Table E.4—continued

2-Digit AFSC	Deployed location	Grade	Gather/Interpret Info	Respect Cultural Diffs	Apply Social Etiquette	Establish Credibility	Change Behavior	Verbal/Nonverbal Comm	Manage Stress	Negotiate with Others	Apply Regional Knowledge	Resolve Conflict	Self-Initiated Learning	Foreign Lang Skills	Influence Others	Establish Authority	# Respondents	Invited Sample	Population
Paralegal (5JXXX)	IRAQ/AFG	LO															**0**	1	1
		HI	.48	.67	.77	.52	.67	.44	.74	.37	.63	.41	.74	.30	.33	.33	**26-27**	67	67
	OTHER	LO															**0-1**	6	6
		HI	.75	1.00	1.00	.67	.83	.75	.83	.67	.83	.67	.92	.42	.67	.67	**12**	44	44
Chaplain Assistant (5RXXX)	IRAQ/AFG	LO															**3**	24	24
		HI	.57	.71	.79	.57	.71	.50	.86	.50	.71	.50	.79	.29	.57	.50	**14**	48	48
	OTHER	LO															**3-4**	21	21
		HI	.92	.92	1.00	.69	1.00	.85	.85	.62	.92	.69	.92	.46	.69	.69	**13**	33	33
Contracting (6CXXX)	IRAQ/AFG	LO	.90	.90	.80	.80	.80	.50	.90	.90	.80	.70	.90	.40	.70	.70	**10**	51	51
		HI	.77	.81	.86	.81	.79	.57	.74	.86	.79	.69	.88	.33	.76	.74	**57-58**	150	211
	OTHER	LO	1.00	.86	1.00	.86	1.00	.57	1.00	1.00	1.00	.86	1.00	.57	.71	.86	**7**	59	59
		HI	.73	.84	.78	.82	.76	.61	.78	.86	.80	.67	.86	.29	.71	.73	**51**	150	168
Financial (6FXXX)	IRAQ/AFG	LO	.60	.60	.60	.60	.60	.60	.60	.60	.60	.60	.80	.60	.60	.60	**5**	55	55
		HI	.63	.79	.58	.58	.67	.54	.79	.54	.57	.50	.79	.50	.46	.54	**23-24**	118	118
	OTHER	LO															**4**	103	103
		HI	.69	.79	.79	.71	.76	.60	.76	.69	.76	.67	.84	.40	.62	.63	**41-43**	150	156
Special Investigations (OSI) (7SXXX)	IRAQ/AFG	LO															**0**	0	0
		HI	.83	.83	.78	.83	.83	.58	.79	.79	.83	.79	.83	.50	.83	.79	**23-24**	121	121
	OTHER	LO															**0**	0	0
		HI	1.00	1.00	1.00	.88	1.00	.75	1.00	.88	1.00	.88	1.00	.63	.88	.75	**8**	81	81
Enlisted Aide (8AXXX)	IRAQ/AFG	LO															**0**	0	0
		HI															**0**	1	1
	OTHER	LO															**0**	0	0
		HI															**0**	9	9
Military Training (8BXXX)	IRAQ/AFG	LO															**0**	1	1
		HI															**4**	16	16
	OTHER	LO															**0**	0	0
		HI															**4**	15	15
Family Support Center (8CXXX)	IRAQ/AFG	LO															**0**	0	0
		HI															**0**	0	0
	OTHER	LO															**0**	0	0
		HI															**1**	6	6
Linguist Debriefer (8DXXX)	IRAQ/AFG	LO															**0**	0	0
		HI	.63	.63	.63	.50	.63	.63	.63	.50	.63	.50	.63	.38	.50	.57	**7-8**	11	11
	OTHER	LO															**0**	0	0
		HI															**0**	0	0
First Sergeant (8FXXX)	IRAQ/AFG	LO															**0**	0	0
		HI	.58	.70	.65	.53	.68	.43	.74	.51	.68	.51	.74	.40	.53	.51	**52-53**	141	141
	OTHER	LO															**0**	0	0

RAND *MG811-T.E.4e*

Table E.4—continued

2-Digit AFSC	Deployed location	Grade	Gather/Interpret Info	Respect Cultural Diffs	Apply Social Etiquette	Establish Credibility	Change Behavior	Verbal/Nonverbal Comm	Manage Stress	Negotiate with Others	Apply Regional Knowledge	Resolve Conflict	Self-Initiated Learning	Foreign Lang Skills	Influence Others	Establish Authority	# Respondents	Invited Sample	Population
USAF Honor Guard (8GXXX)	IRAQ/AFG	LO															0	0	0
		HI															0	1	1
	OTHER	LO															0	0	0
		HI															0	0	0
Postal Specialist (8MXXX)	IRAQ/AFG	LO															0	22	22
		HI	.80	.70	.70	.70	.80	.70	.90	.70	.89	.70	.80	.40	.70	.00	9-10	52	52
	OTHER	LO	.69	.77	.77	.69	.69	.62	.77	.62	.77	.62	.85	.46	.77	.77	13	89	89
		HI	.66	.74	.63	.63	.71	.53	.74	.59	.68	.58	.79	.42	.53	.55	37-38	100	100
Recruiter (8RXXX)	IRAQ/AFG	LO															0	3	3
		HI															0-1	6	6
	OTHER	LO															0	4	4
		HI															0-3	15	15
Professional Military Education Instructor (8TXXX)	IRAQ/AFG	LO															1	0	0
		HI															0	1	1
	OTHER	LO															0	0	0
		HI															0	3	3
Awaiting Retraining (9AXXX)	IRAQ/AFG	LO															1	1	1
		HI															0	0	0
	OTHER	LO															0	4	4
		HI															0	0	0
Dorm Manager (9DXXX)	IRAQ/AFG	LO															0	0	0
		HI															0	9	9
	OTHER	LO															0	1	1
		HI															0	10	10
Senior Enlisted Advisor (9EXXX)	IRAQ/AFG	LO															0	0	0
		HI															0	1	1
	OTHER	LO															0	0	0
		HI															0	4	4
Group Superintendent (9GXXX)	IRAQ/AFG	LO															0	0	0
		HI	.80	.90	.70	.80	.80	.70	.90	.70	.80	.80	.90	.60	.70	.70	10	25	25
	OTHER	LO															0	0	0
		HI	.60	.90	.80	.60	.70	.50	.80	.50	.78	.50	.70	.40	.50	.50	9-10	23	23
Interpreter / Translator (9LXXX)	IRAQ/AFG	LO															0	0	0
		HI															0	0	0
	OTHER	LO															0	0	0
		HI															0	3	3
Technical Applications Specialist (9SXXX)	IRAQ/AFG	LO															0	0	0
		HI															0-1	0	0
	OTHER	LO															0	5	5
		HI	.40	.40	.40	.20	.40	.20	.40	.20	.60	.20	.60	.20	.20	.20	5	18	18

RAND *MG811-T.E.4f*

Table E.4—continued

2-Digit AFSC	Deployed location	Grade	Gather/Interpret Info	Respect Cultural Diffs	Apply Social Etiquette	Establish Credibility	Change Behavior	Verbal/Nonverbal Comm	Manage Stress	Negotiate with Others	Apply Regional Knowledge	Resolve Conflict	Self-Initiated Learning	Foreign Lang Skills	Influence Others	Establish Authority	# Respondents	Invited Sample	Population
Basic Enlisted Trainee (9TXXX)	IRAQ/AFG	LO															0	13	13
		HI															0	0	0
	OTHER	LO															0-1	8	8
		HI															0-1	0	0
Unallotted Airman (9UXXX)	IRAQ/AFG	LO	.62	.86	.50	.64	.57	.43	.71	.54	.71	.64	.57	.36	.64	.64	13-14	0	0
		HI	.63	.69	.69	.63	.69	.69	.69	.50	.63	.63	.69	.31	.69	.69	16	0	0
	OTHER	LO	.86	.86	.79	.79	.86	.71	.93	.79	.86	.79	.93	.43	.79	.79	14	0	0
		HI	.67	.78	.56	.78	.78	.44	.78	.78	.67	.78	.89	.33	.78	.78	9	0	0
TCN ESCORT	IRAQ/AFG	LO	.54	.71	.67	.63	.67	.42	.71	.58	.65	.58	.61	.29	.63	.67	23-24	0	0
		HI	.61	.87	.78	.57	.65	.48	.70	.48	.70	.61	.83	.35	.52	.57	23	0	0
	OTHER	LO	.78	.85	.73	.83	.78	.55	.80	.80	.80	.83	.85	.46	.83	.85	40-41	0	0
		HI	.77	.77	.81	.81	.81	.65	.81	.65	.81	.68	.81	.58	.69	.73	25-26	0	0

RAND *MG811-T.E.4g*

Table E.5
Officer Predeployment Training, Proportion Reporting Being Trained, by AFSC, Grade, and Deployment Subgroups

2-Digit AFSC	Deployed Location	Grade	Gather/Interpret Info	Respect Cultural Diffs	Apply Social Etiquette	Establish Credibility	Change Behavior	Verbal/Nonverbal Comm	Manage Stress	Negotiate with Others	Apply Regional Knowledge	Resolve Conflict	Self-Initiated Learning	Foreign Lang Skills	Influence Others	Establish Authority	# Respondents	Invited Sample	Population
All Deployed Personnel	All Deployed Personnel	All Deployed Personnel	.59	.72	.65	.56	.64	.45	.65	.52	.67	.53	.75	.39	.53	.54	6,045-6,075	23,020	88,304
Commander (10XX)	IRAQ/AFG	LO															0	0	0
		HI															4	18	18
	OTHER	LO															0	0	0
		HI	.63	.75	.63	.50	.50	.50	.50	.38	.75	.38	.75	.50	.38	.50	8	32	32
Pilot (11XX)	IRAQ/AFG	LO	.56	.66	.69	.47	.56	.34	.59	.41	.74	.38	.72	.31	.44	.44	31-32	150	700
		HI	.54	.65	.59	.52	.59	.46	.57	.48	.63	.52	.63	.43	.56	.52	62-63	150	392
	OTHER	LO	.66	.75	.81	.69	.75	.63	.78	.66	.81	.69	.72	.56	.66	.63	32	150	2,711
		HI	.54	.68	.61	.51	.63	.44	.58	.47	.58	.51	.68	.35	.48	.49	56-57	150	962
Navigator (12XX)	IRAQ/AFG	LO	.59	.79	.68	.62	.74	.56	.68	.56	.85	.56	.79	.56	.56	.56	34	150	254
		HI	.37	.45	.34	.24	.34	.26	.34	.24	.58	.24	.47	.26	.29	.29	37-38	150	161
	OTHER	LO	.70	.73	.73	.57	.67	.40	.60	.47	.77	.53	.73	.40	.53	.53	30	150	900
		HI	.57	.72	.65	.53	.67	.54	.61	.53	.72	.53	.69	.46	.54	.54	71-72	150	372
Space, Missile, and Command and Control (13XX)	IRAQ/AFG	LO	.65	.74	.71	.58	.65	.52	.68	.55	.81	.61	.77	.45	.55	.58	31	150	153
		HI	.41	.68	.59	.45	.45	.36	.50	.45	.55	.45	.73	.36	.45	.45	22	70	70
	OTHER	LO	.52	.76	.69	.52	.62	.41	.59	.41	.66	.45	.72	.41	.45	.45	29	150	413
		HI	.47	.63	.59	.42	.50	.31	.53	.39	.61	.36	.64	.23	.34	.38	64	150	184
Intelligence (14XX)	IRAQ/AFG	LO	.57	.64	.52	.42	.56	.38	.56	.36	.68	.36	.80	.26	.38	.38	49-50	150	236
		HI	.54	.54	.38	.38	.46	.38	.46	.38	.58	.33	.83	.29	.42	.42	24	68	68
	OTHER	LO	.58	.70	.67	.49	.67	.40	.60	.40	.67	.42	.72	.30	.44	.44	43	150	427
		HI	.62	.85	.69	.46	.69	.36	.64	.44	.74	.41	.82	.33	.44	.38	38-39	108	108
Weather (15XX)	IRAQ/AFG	LO	.67	.83	.67	.33	.67	.33	.50	.33	.67	.50	.67	.33	.33	.50	6	37	37
		HI	.57	.71	.71	.57	.71	.57	.71	.57	.71	.57	.86	.57	.57	.57	7	15	15
	OTHER	LO	.45	.64	.55	.64	.55	.27	.55	.64	.73	.64	1.00	.27	.64	.55	11	38	38
		HI															0-4	9	9
International Affairs/FAO (16XX)	IRAQ/AFG	LO	.46	.54	.62	.31	.38	.31	.54	.31	.31	.31	.62	.31	.31	.31	13	6	6
		HI	.34	.51	.49	.36	.38	.26	.43	.30	.45	.34	.70	.23	.30	.36	47	39	39
	OTHER	LO	.67	.67	.67	.67	.67	.67	.67	.67	.67	.67	.83	.67	.67	.67	6	5	5
		HI	.33	.56	.39	.33	.39	.39	.39	.33	.44	.33	.72	.33	.33	.33	18	43	43
Logistics Commander (20XX)	IRAQ/AFG	LO															0	0	0
		HI															4	18	18
	OTHER	LO															0	0	0
		HI	.67	.83	.67	.67	.67	.50	.67	.50	.67	.50	.67	.50	.50	.50	6	10	10
Aircraft Maintenance (21XX)	IRAQ/AFG	LO	.33	.36	.42	.33	.34	.31	.39	.33	.33	.33	.42	.28	.36	.36	35-36	150	295
		HI	.48	.63	.56	.48	.54	.43	.59	.43	.63	.43	.67	.35	.44	.45	53-54	133	133
	OTHER	LO	.56	.64	.56	.42	.64	.33	.50	.39	.66	.39	.67	.33	.42	.44	35-36	150	470
		HI	.51	.60	.57	.51	.50	.45	.49	.47	.53	.47	.72	.45	.49	.47	46-47	150	153

RAND *MG811-T.E.5a*

Table E.5—continued

2-Digit AFSC	Deployed Location	Grade	Gather/Interpret Info	Respect Cultural Diffs	Apply Social Etiquette	Establish Credibility	Change Behavior	Verbal/Nonverbal Comm	Manage Stress	Negotiate with Others	Apply Regional Knowledge	Resolve Conflict	Self-Initiated Learning	Foreign Lang Skills	Influence Othe's	Establish Authority	# Respondents	Invited Sample	Population
Support Commander (30XX)	IRAQ/AFG	LO															0	0	0
		HI	.30	.40	.40	.30	.40	.20	.40	.30	.40	.30	.80	.20	.30	.30	10	26	26
	OTHER	LO															0	0	0
		HI	.67	.89	.78	.78	.89	.67	.89	.67	.89	.67	.89	.67	.78	.67	9	26	26
Security Forces (31XX)	IRAQ/AFG	LO	.68	.77	.71	.61	.68	.60	.65	.55	.77	.58	.77	.45	.68	.65	30-31	150	178
		HI	.54	.69	.62	.54	.46	.31	.62	.54	.54	.54	.75	.23	.62	.62	12-13	45	45
	OTHER	LO	.67	.78	.67	.78	.67	.44	.56	.67	.67	.56	.89	.33	.56	.56	9	50	50
		HI															0-1	12	12
Civil Engineer (32XX)	IRAQ/AFG	LO	.71	.76	.79	.62	.76	.58	.71	.62	.82	.59	.79	.41	.62	.62	33-34	150	218
		HI	.60	.71	.64	.62	.62	.45	.62	.57	.62	.62	.79	.43	.60	.62	42	98	98
	OTHER	LO	.54	.71	.63	.54	.54	.42	.54	.54	.50	.50	.79	.38	.50	.46	24	139	139
		HI	.53	.80	.80	.53	.73	.40	.67	.33	.47	.47	.67	.40	.33	.33	15	48	48
Comm-Information Systems (33XX)	IRAQ/AFG	LO	.60	.67	.58	.53	.56	.42	.60	.49	.60	.53	.70	.37	.51	.51	43	150	317
		HI	.43	.57	.45	.45	.50	.36	.52	.39	.57	.41	.63	.36	.39	.41	43-44	91	91
	OTHER	LO	.37	.63	.53	.37	.63	.23	.57	.30	.53	.30	.69	.23	.30	.33	29-30	150	397
		HI	.56	.62	.54	.48	.54	.42	.54	.44	.60	.48	.63	.38	.44	.44	52	94	94
Services (34XX)	IRAQ/AFG	LO	.50	.60	.50	.40	.50	.30	.45	.36	.45	.35	.70	.30	.35	.45	20	45	45
		HI															3	15	15
	OTHER	LO	.61	.79	.63	.63	.58	.50	.58	.53	.68	.50	.74	.42	.63	.58	18-19	87	87
		HI	.86	.86	.86	.71	.86	.86	.71	.71	.86	.71	.86	.71	.71	.71	7	18	18
Public Affairs (35XX)	IRAQ/AFG	LO	.50	.67	.50	.50	.50	.50	.50	.50	.67	.50	.67	.50	.50	.50	6	32	32
		HI	.57	.71	.57	.43	.43	.29	.43	.43	.43	.29	.57	.14	.57	.57	7	11	11
	OTHER	LO	.80	.80	.60	.70	.80	.70	.60	.60	.80	.60	.80	.60	.70	.60	10	40	40
		HI															0-5	25	25
Personnel (36XX)	IRAQ/AFG	LO															0	26	26
		HI															0	15	15
	OTHER	LO															0	41	41
		HI															0	9	9
Manpower-Personnel (37XX)	IRAQ/AFG	LO	.44	.50	.44	.39	.56	.33	.56	.39	.44	.39	.56	.33	.39	.39	18	51	51
		HI	.22	.39	.50	.22	.35	.22	.35	.22	.30	.22	.61	.17	.22	.22	22-23	36	36
	OTHER	LO	.71	.74	.60	.57	.71	.45	.69	.48	.69	.55	.83	.43	.52	.60	42	83	83
		HI	.50	.71	.57	.31	.64	.36	.64	.31	.57	.31	.64	.36	.38	.46	13-14	31	31
Manpower (38XX)	IRAQ/AFG	LO															0	3	3
		HI															0	1	1
	OTHER	LO															0	13	13
		HI															0	3	3

RAND *MG811-T.E.5b*

Table E.5—continued

2-Digit AFSC	Deployed Location	Grade	Gather/Interpret Info	Respect Cultural Diffs	Apply Social Etiquette	Establish Credibility	Change Behavior	Verbal/Nonverbal Comm	Manage Stress	Negotiate with Others	Apply Regional Knowledge	Resolve Conflict	Self-Initiated Learning	Foreign Lang Skills	Influence Others	Establish Authority	# Respondents	Invited Sample	Population
Medical Commander (40XX)	IRAQ/AFG	LO															**0**	0	0
		HI	.60	.80	.60	.80	1.00	.40	1.00	.80	.80	.80	1.00	.40	.80	.80	**5**	7	7
	OTHER	LO															**0**	0	0
		HI	.29	.57	.57	.43	.43	.29	.43	.29	.29	.43	.43	.29	.29	.29	**7**	14	14
Health Services Administrator (41XX)	IRAQ/AFG	LO	.33	.56	.56	.33	.44	.33	.44	.33	.44	.33	.67	.33	.33	.33	**9**	40	40
		HI	.69	.88	.81	.75	.88	.63	.81	.69	.81	.75	.94	.63	.75	.76	**16**	44	44
	OTHER	LO	.63	.50	.38	.50	.50	.38	.50	.50	.63	.50	1.00	.38	.38	.38	**8**	44	44
		HI	.25	.67	.42	.25	.33	.25	.67	.25	.58	.25	.58	.25	.25	.33	**12**	37	37
Biomedical Clinician (42XX)	IRAQ/AFG	LO	.64	.86	.73	.64	.64	.55	.77	.59	.68	.62	.77	.50	.64	.55	**21-22**	70	70
		HI	.56	.67	.56	.56	.56	.44	.56	.44	.56	.44	.89	.33	.44	.44	**9**	34	34
	OTHER	LO															**3**	48	48
		HI															**4**	17	17
Biomedical Specialists (43XX)	IRAQ/AFG	LO	.50	.50	.38	.38	.50	.38	.50	.38	.50	.38	.50	.38	.38	.38	**8**	32	32
		HI	.57	.62	.67	.43	.62	.43	.62	.38	.71	.43	.81	.33	.48	.38	**21**	49	49
	OTHER	LO	.69	.77	.77	.46	.69	.23	.77	.38	.69	.38	.85	.15	.38	.38	**13**	34	34
		HI	.48	.68	.59	.45	.55	.36	.50	.36	.55	.27	.68	.23	.32	.32	**21-22**	42	42
Physician (44XX)	IRAQ/AFG	LO															**4**	46	46
		HI	.53	.63	.56	.47	.53	.38	.53	.44	.50	.41	.63	.31	.44	.44	**32**	120	120
	OTHER	LO															**0-3**	48	48
		HI	.70	.83	.65	.61	.70	.52	.65	.57	.74	.55	.78	.43	.61	.57	**22-23**	77	77
Surgery (45XX)	IRAQ/AFG	LO															**0-1**	16	16
		HI	.44	.78	.67	.61	.56	.39	.61	.44	.67	.50	.72	.33	.61	.56	**18**	113	113
	OTHER	LO															**0**	4	4
		HI	.75	.75	.75	.75	.75	.75	.75	.63	.88	.75	.88	.63	.63	.75	**8**	38	38
Nurse (46XX)	IRAQ/AFG	LO	.47	.67	.57	.49	.51	.33	.53	.39	.52	.42	.65	.31	.39	.38	**50-51**	150	258
		HI	.62	.69	.70	.52	.65	.54	.71	.50	.71	.48	.77	.50	.54	.54	**47-48**	99	99
	OTHER	LO	.53	.66	.60	.53	.57	.55	.55	.51	.60	.55	.66	.47	.57	.57	**47**	150	269
		HI	.64	.79	.78	.67	.69	.61	.64	.67	.79	.67	.91	.58	.67	.67	**32-33**	94	94
Dental (47XX)	IRAQ/AFG	LO															**0**	4	4
		HI															**0-4**	12	12
	OTHER	LO															**0-1**	8	8
		HI	.45	.64	.55	.45	.55	.45	.55	.45	.55	.45	.73	.36	.55	.45	**11**	24	24
Aerospace Medicine (48XX)	IRAQ/AFG	LO															**0-4**	26	26
		HI	.65	.76	.82	.65	.82	.41	.71	.53	.65	.53	.76	.35	.53	.53	**17**	51	51
	OTHER	LO	.50	.50	.50	.33	.50	.33	.67	.33	.50	.33	.50	.33	.33	.33	**6**	57	57
		HI	.65	.71	.71	.58	.71	.58	.68	.58	.65	.53	.61	.42	.55	.52	**30-31**	81	81

RAND *MG811-T.E.5c*

Table E.5—continued

2-Digit AFSC	Deployed Location	Grade	Gather/Interpret Info	Respect Cultural Diffs	Apply Social Etiquette	Establish Credibility	Change Behavior	Verbal/Nonverbal Comm	Manage Stress	Negotiate with Others	Apply Regional Knowledge	Resolve Conflict	Self-Initiated Learning	Foreign Lang Skills	Influence Others	Establish Authority	# Respondents	Invited Sample	Population
Law (51XX)	IRAQ/AFG	LO	.88	.88	.75	.75	.63	.38	.75	.50	.88	.63	.88	.38	.63	.75	8	56	56
		HI	.57	.76	.62	.57	.62	.38	.57	.52	.71	.57	.71	.33	.52	.48	21	43	43
	OTHER	LO															4	37	37
		HI	.69	.75	.75	.63	.69	.63	.81	.69	.75	.63	.88	.56	.69	.69	16	49	49
Chaplain (52XX)	IRAQ/AFG	LO	.60	.67	.73	.60	.73	.40	.73	.53	.53	.47	.93	.40	.60	.53	15	43	43
		HI	.53	.79	.68	.58	.74	.26	.63	.47	.58	.47	.95	.26	.53	.42	19	49	49
	OTHER	LO	.91	1.00	.80	.73	1.00	.55	.82	.55	.82	.45	.91	.45	.45	.45	10-11	43	43
		HI	.40	.80	.73	.50	.80	.36	.50	.50	.60	.30	.90	.36	.40	.50	10-11	25	25
Program Director (60XX)	IRAQ/AFG	LO															0	0	0
		HI															0	1	1
	OTHER	LO															0	0	0
		HI															0	0	0
Scientific/Research (61XX)	IRAQ/AFG	LO															0-3	9	9
		HI															0-4	4	4
	OTHER	LO	.67	.83	.67	.67	.67	.67	.67	.67	.67	.67	.83	.67	.67	.67	6	15	15
		HI	.29	.43	.43	.29	.43	.29	.43	.29	.29	.29	.29	.29	.29	.29	7	16	16
Developmental Engineering (62XX)	IRAQ/AFG	LO	.33	.50	.50	.33	.33	.17	.33	.33	.50	.33	.50	.17	.33	.33	6	21	21
		HI															0-2	10	10
	OTHER	LO	.71	.71	.86	.71	.71	.71	.71	.71	.71	.71	.86	.71	.71	.71	7	31	31
		HI															0	5	5
Acquisition (63XX)	IRAQ/AFG	LO	.63	.75	.63	.50	.75	.50	.50	.50	.75	.50	.63	.38	.50	.50	8	33	33
		HI	.67	.67	.67	.67	.67	.44	.67	.56	.67	.56	.78	.44	.56	.44	9	42	42
	OTHER	LO															2	31	31
		HI															3	17	17
Contracting (64XX)	IRAQ/AFG	LO	.56	.69	.63	.63	.66	.44	.63	.68	.59	.56	.72	.38	.63	.56	31-32	111	111
		HI	.57	.81	.67	.67	.62	.52	.57	.65	.62	.48	.76	.33	.52	.48	20-21	52	52
	OTHER	LO	.50	.77	.59	.68	77	.27	.68	.68	.68	.55	.77	.24	.62	.50	21-22	69	69
		HI															3	4	4
Finance (65XX)	IRAQ/AFG	LO	.50	.88	.75	.50	.63	.38	.75	.25	.75	.50	.75	.25	.50	.50	8	43	43
		HI	.60	.60	.60	.60	.60	.40	.60	.60	.60	.60	.80	.40	.60	.60	5	37	37
	OTHER	LO	.60	.40	.40	.40	.40	.20	.20	.40	.60	.20	.60	.20	.40	.40	5	22	22
		HI	.63	.88	.63	.88	.88	.44	.75	.75	.63	.75	.75	.44	.75	.75	8-9	17	17
Special Investigations (71XX)	IRAQ/AFG	LO															4	30	30
		HI															4	14	14
	OTHER	LO															3-4	17	17
		HI															0-1	4	4

RAND *MG811-T.E.5d*

Table E.5—continued

2-Digit AFSC	Deployed Location	Grade	Gather/Interpret Info	Respect Cultural Diffs	Apply Social Etiquette	Establish Credibility	Change Behavior	Verbal/Nonverbal Comm	Manage Stress	Negotiate with Others	Apply Regional Knowledge	Resolve Conflict	Self-Initiated Learning	Foreign Lang Skills	Influence Others	Establish Authority	# Respondents	Invited Sample	Population
Commander—USAFA Cadet Squadron (80XX)	IRAQ/AFG	LO															0	0	0
		HI															0	0	0
	OTHER	LO															0	0	0
		HI															0	1	1
Instructor (81XX)	IRAQ/AFG	LO															1	4	4
		HI															0-3	8	8
	OTHER	LO															0	6	6
		HI															0	8	8
Academic Program Manager (82XX)	IRAQ/AFG	LO															0	1	1
		HI															0	3	3
	OTHER	LO															0	1	1
		HI															0	3	3
Recruiting Service (83XX)	IRAQ/AFG	LO															0	4	4
		HI															0	0	0
	OTHER	LO															0	2	2
		HI															0	0	0
Historian (84XX)	IRAQ/AFG	LO															0	0	0
		HI															0	0	0
	OTHER	LO															0	0	0
		HI															0	0	0
USAF Honor Guard (85XX)	IRAQ/AFG	LO															0	0	0
		HI															0	0	0
	OTHER	LO															0	0	0
		HI															0	1	1
Command and Control (86XX)	IRAQ/AFG	LO															0	2	2
		HI															2-3	3	3
	OTHER	LO															3	3	3
		HI															0	5	5
Inspector General (87XX)	IRAQ/AFG	LO															0	0	0
		HI															0	0	0
	OTHER	LO															0	0	0
		HI															0	1	1
Protocol/Aide (88XX)	IRAQ/AFG	LO															0-2	1	1
		HI															0	0	0
	OTHER	LO															3	0	0
		HI															0	0	0

RAND *MG811-T.E.5e*

Table E.5—continued

2-Digit AFSC	Deployed Location	Grade	Gather/Interpret Info	Respect Cultural Diffs	Apply Social Etiquette	Establish Credibility	Change Behavior	Verbal/Nonverbal Comm	Manage Stress	Negotiate with Others	Apply Regional Knowledge	Resolve Conflict	Self-Initiated Learning	Foreign Lang Skills	Influence Others	Establish Authority	# Respondents	Invited Sample	Population
General Officer (90XX)	IRAQ/AFG	LO															0	0	0
		HI															3	10	10
	OTHER	LO															0	0	0
		HI															0	11	11
Commander (91XX)	IRAQ/AFG	LO															0	0	0
		HI															4	13	13
	OTHER	LO															0	0	0
		HI	.56	.56	.67	.56	.67	.44	.67	.56	.56	.56	.67	.44	.56	.56	9	18	18
Student (92XX)	IRAQ/AFG	LO															0	22	22
		HI															0	7	7
	OTHER	LO															0-2	65	65
		HI															0	12	12
Patient (93XX)	IRAQ/AFG	LO															0	0	0
		HI															0	0	0
	OTHER	LO															0	0	0
		HI															0	0	0
Nuclear Weapons Custodian (94XX)	IRAQ/AFG	LO															0	0	0
		HI															0	0	0
	OTHER	LO															0	0	0
		HI															0	0	0
Non-extended Active Duty (95XX)	IRAQ/AFG	LO															0	0	0
		HI															0	0	0
	OTHER	LO															0	0	0
		HI															0	1	1
Unclassified Officer (96XX)	IRAQ/AFG	LO															0	0	0
		HI															0	0	0
	OTHER	LO															0-1	2	2
		HI															0	0	0
Executive Officer above Wing Level (97XX)	IRAQ/AFG	LO															1	3	3
		HI															0-2	12	12
	OTHER	LO															0	6	6
		HI															0	7	7

RAND *MG811-T.E.5f*

Table E.6
Enlisted Predeployment Training, Proportion Reporting Being Trained, by AFSC, Grade, and Deployment Subgroups

2-Digit AFSC	Deployed Location	Grade	Gather/Interpret Info	Respect Cultural Diffs	Apply Social Etiquette	Establish Credibility	Change Behavior	Verbal/Nonverbal Comm	Manage Stress	Negotiate with Others	Apply Regional Knowledge	Resolve Conflict	Self-Initiated Learning	Foreign Lang Skills	Influence Others	Establish Authority	# Respondents	Invited Sample	Population
All Deployed Personnel	All Deployed Personnel	All Deployed Personnel	.59	.72	.65	.56	.64	.45	.65	.52	.67	.53	.75	.39	.53	.54	6,045-6,075	23,020	88,304
Aircrew Operations (1AXXX)	IRAQ/AFG	LO	.80	.80	.87	.73	.87	.60	.87	.67	1.00	.67	.93	.60	.73	.73	15	150	218
		HI	.36	.59	.59	.36	.56	.31	.54	.36	.56	.33	.62	.26	.38	.38	39	150	633
	OTHER	LO	.83	.83	.83	.75	.88	.63	.79	.71	.83	.75	.83	.50	.67	.71	24	150	1,248
		HI	.69	.74	.71	.60	.71	.45	.69	.50	.79	.55	.86	.40	.55	.00	41-42	150	1,431
Command & Control Systems Operations (1CXXX)	IRAQ/AFG	LO	.63	.84	.68	.74	.79	.53	.74	.63	.68	.68	.79	.53	.63	.63	19	150	859
		HI	.62	.78	.72	.62	.69	.59	.74	.60	.70	.62	.76	.51	.64	.66	49-51	150	983
	OTHER	LO	.85	.90	.85	.80	.85	.65	.84	.80	.85	.65	.90	.55	.75	.65	19-20	150	554
		HI	.60	.71	.65	.59	.68	.48	.71	.52	.73	.58	.75	.41	.53	.57	62-63	150	676
Intelligence (1NXXX)	IRAQ/AFG	LO	.73	.73	.80	.73	.73	.67	.80	.67	.93	.73	.93	.53	.73	.80	15	150	290
		HI	.60	.63	.58	.42	.53	.33	.54	.40	.61	.37	.81	.32	.40	.40	57	150	313
	OTHER	LO	.63	.85	.70	.30	.65	.40	.60	.35	.85	.30	.75	.25	.35	.35	19-20	150	436
		HI	.63	.80	.69	.55	.69	.43	.73	.49	.71	.52	.84	.41	.55	.59	48-49	150	437
Safety (1SXXX)	IRAQ/AFG	LO															0	0	0
		HI	.33	.67	.47	.33	.27	.20	.33	.27	.53	.27	.73	.13	.27	.33	15	39	39
	OTHER	LO															0	1	1
		HI	.43	.64	.64	.29	.43	.14	.36	.21	.50	.21	.50	.14	.29	.36	14	41	41
Aircrew Protection (1TXXX)	IRAQ/AFG	LO	.67	.67	.67	.56	.56	.56	.56	.56	.56	.56	.56	.56	.56	.56	9	115	115
		HI	.48	.63	.56	.44	.59	.33	.59	.41	.63	.44	.63	.33	.41	.41	27	118	118
	OTHER	LO	.67	.67	.75	.67	.67	.50	.67	.58	.67	.58	.67	.42	.58	.58	12	150	276
		HI	.54	.67	.61	.49	.67	.37	.67	.41	.61	.46	.76	.37	.46	.46	45-46	150	290
Weather (1WXXX)	IRAQ/AFG	LO	.50	.83	.67	.67	.67	.50	.50	.67	.67	.67	.83	.50	.67	.67	6	68	68
		HI	.55	.73	.65	.50	.65	.43	.58	.47	.72	.46	.72	.33	.48	.50	58-60	150	217
	OTHER	LO															4	54	54
		HI	.52	.73	.64	.45	.66	.41	.66	.41	.67	.42	.80	.36	.41	.43	43-44	127	127
Manned Aerospace Maintenance (2AXXX)	IRAQ/AFG	LO	.50	.60	.60	.50	.60	.10	.60	.30	.50	.56	.70	.10	.40	.30	9-10	150	2,505
		HI	.54	.70	.65	.49	.60	.39	.68	.43	.63	.46	.71	.30	.48	.52	62-63	150	2,396
	OTHER	LO	.50	.83	.75	.75	.75	.50	.75	.67	.67	.67	.83	.33	.75	.67	12	150	5,727
		HI	.54	.65	.67	.54	.60	.47	.67	.53	.63	.54	.70	.47	.51	.53	57	150	5,923
Communications & Electronics (2EXXX)	IRAQ/AFG	LO	.74	.83	.74	.70	.74	.57	.87	.61	.83	.65	.91	.48	.65	.70	23	150	705
		HI	.47	.69	.59	.47	.55	.42	.60	.49	.62	.49	.72	.41	.49	.47	77-78	150	596
	OTHER	LO	.58	.81	.73	.42	.74	.26	.81	.35	.71	.39	.71	.23	.35	.42	30-31	150	654
		HI	.65	.72	.72	.60	.68	.47	.72	.53	.65	.53	.74	.42	.56	.56	57	150	689

RAND *MG811-T.E.6a*

Table E.6—continued

2-Digit AFSC	Deployed Location	Grade	Gather/Interpret Info	Respect Cultural Diffs	Apply Social Etiquette	Establish Credibility	Change Behavior	Verbal/Nonverbal Comm	Manage Stress	Negotiate with Others	Apply Regional Knowledge	Resolve Conflict	Self-Initiated Learning	Foreign Lang Skills	Influence Others	Establish Authority	# Respondents	Invited Sample	Population
Fuels (2FXXX)	IRAQ/AFG	LO	.69	.81	.72	.65	.77	.46	.77	.58	.81	.65	.81	.42	.69	.69	**25-26**	150	209
		HI	.58	.75	.60	.55	.65	.42	.68	.53	.62	.52	.73	.37	.55	.55	**59-60**	150	264
	OTHER	LO	.73	.87	.80	.60	.80	.67	.73	.67	.71	.60	.73	.53	.73	.73	**14-15**	150	668
		HI	.65	.77	.74	.62	.72	.48	.71	.52	.72	.55	.75	.39	.52	.52	**69**	150	709
Logistics Plans (2GXXX)	IRAQ/AFG	LO															**0-2**	22	22
		HI	.66	.69	.66	.61	.71	.55	.69	.62	.66	.62	.72	.55	.62	.62	**28-29**	87	87
	OTHER	LO	.86	.86	.86	.71	.86	.57	.86	.71	.83	.71	.86	.43	.71	.57	**6-7**	37	37
		HI	.51	.69	.61	.49	.59	.44	.72	.46	.67	.49	.77	.36	.46	.51	**38-39**	119	119
Missile & Space Systems Maintenance (2MXXX)	IRAQ/AFG	LO															**0-3**	59	59
		HI															**4**	22	22
	OTHER	LO															**0**	39	39
		HI															**1**	7	7
Precision Measurement Equipment Lab (2PXXX)	IRAQ/AFG	LO															**0**	5	5
		HI															**0-2**	5	5
	OTHER	LO	.60	.60	.60	.40	.60	.40	.60	.20	.60	.20	.60	.20	.40	.40	**5**	42	42
		HI	.48	.68	.60	.39	.68	.39	.65	.37	.52	.39	.61	.39	.35	.39	**30-31**	61	61
Maintenance Management Systems (2RXXX)	IRAQ/AFG	LO															**4**	42	42
		HI	.80	1.00	.70	.60	.90	.30	.80	.40	.60	.50	.80	.20	.70	.70	**10**	35	35
	OTHER	LO	.80	1.00	.67	.67	.83	.33	1.00	.50	.67	.67	.83	.33	.67	.50	**5-6**	92	92
		HI	.78	.78	.89	.67	.85	.56	.85	.63	.74	.67	.78	.48	.63	.63	**27**	100	100
Supply (2SXXX)	IRAQ/AFG	LO	.82	1.00	.91	.82	1.00	.91	1.00	.82	.91	.82	.91	.64	.82	.82	**11**	150	374
		HI	.53	.65	.61	.47	.59	.43	.65	.45	.63	.45	.63	.35	.45	.45	**47-49**	150	643
	OTHER	LO	.85	.85	.62	.85	.77	.54	.85	.69	.85	.77	.77	.46	.62	.62	**13**	150	650
		HI	.58	.72	.68	.64	.72	.51	.77	.59	.69	.62	.74	.49	.59	.62	**38-39**	150	1,067
Transportation & Vehicle Maintenance (2TXXX)	IRAQ/AFG	LO	.75	.83	.71	.75	.71	.58	.83	.75	.75	.75	.79	.54	.75	.75	**24**	150	1,232
		HI	.69	.76	.69	.62	.66	.51	.69	.60	.76	.63	.79	.46	.62	.68	**67-68**	150	1,219
	OTHER	LO	.71	.79	.71	.71	.79	.04	.79	.64	.71	.64	.79	.50	.71	.71	**14**	150	1,207
		HI	.74	.81	.79	.64	.78	.49	.68	.60	.74	.59	.87	.47	.64	.77	**45-47**	150	1,275
Munitions & Weapons (2WXXX)	IRAQ/AFG	LO	.65	.83	.74	.70	.78	.52	.83	.61	.74	.70	.70	.52	.61	.61	**23**	150	768
		HI	.51	.61	.53	.45	.55	.43	.65	.45	.63	.46	.61	.33	.43	.45	**50-51**	150	714
	OTHER	LO	.63	.79	.61	.68	.68	.47	.72	.53	.84	.63	.84	.47	.53	.58	**18-19**	150	980
		HI	.56	.74	.66	.52	.65	.47	.67	.49	.61	.53	.56	.45	.51	.53	**56-58**	150	905
Information Management (3AXXX)	IRAQ/AFG	LO	.60	.80	.74	.70	.60	.55	.75	.65	.70	.75	.75	.50	.75	.75	**19-20**	150	220
		HI	.50	.65	.55	.52	.60	.35	.59	.51	.62	.48	.72	.32	.47	.52	**56-60**	150	539
	OTHER	LO	.74	.89	.70	.74	.79	.55	.79	.79	.84	.74	.95	.40	.79	.79	**19-20**	150	329
		HI	.50	.67	.56	.46	.60	.31	.52	.42	.58	.39	.69	.26	.37	.36	**52-54**	150	756

RAND *MG811-T.E.6b*

Table E.6—continued

2-Digit AFSC	Deployed Location	Grade	Gather/Interpret Info	Respect Cultural Diffs	Apply Social Etiquette	Establish Credibility	Change Behavior	Verbal/Nonverbal Comm	Manage Stress	Negotiate with Others	Apply Regional Knowledge	Resolve Conflict	Self-Initiated Learning	Foreign Lang Skills	Influence Others	Establish Authority	# Respondents	Invited Sample	Population
Communications & Computer Systems (3CXXX)	IRAQ/AFG	LO	.70	.78	.78	.65	.74	.57	.78	.71	.70	.70	.74	.52	.70	.70	**21-23**	150	405
		HI	.43	.61	.54	.39	.46	.33	.52	.30	.54	.39	.57	.26	.39	.35	**46**	150	627
	OTHER	LO	.77	.85	.69	.69	.69	.54	.77	.69	.77	.62	.85	.54	.77	.69	**13**	150	617
		HI	.56	.67	.58	.53	.56	.49	.53	.53	.58	.53	.78	.42	.53	.56	**45**	150	1,026
Civil Engineering (3EXXX)	IRAQ/AFG	LO	.78	.78	.83	.61	.83	.56	.83	.44	.78	.72	.67	.50	.50	.50	**18**	150	1,839
		HI	.67	.81	.70	.67	.67	.61	.75	.63	.78	.66	.83	.48	.61	.66	**63-64**	150	1,826
	OTHER	LO	.56	.75	.69	.69	.75	.31	.56	.75	.69	.75	.69	.25	.69	.69	**16**	150	2,039
		HI	.69	.71	.64	.65	.66	.43	.67	.59	.71	.60	.83	.43	.60	.60	**40-42**	150	1,919
Historian (3HXXX)	IRAQ/AFG	LO															**0**	0	0
		HI															**0-1**	5	5
	OTHER	LO															**0**	0	0
		HI															**3**	7	7
Services (3MXXX)	IRAQ/AFG	LO	.74	.91	.74	.74	.78	.52	.83	.65	.65	.61	.96	.30	.70	.70	**23**	150	208
		HI	.61	.84	.73	.64	.70	.48	.73	.55	.73	.61	.80	.41	.60	.66	**43-44**	150	268
	OTHER	LO	.73	.73	.80	.80	.80	.60	.93	.60	.80	.73	.80	.40	.80	.73	**15**	150	703
		HI	.77	.85	.83	.77	.80	.58	.87	.76	.81	.71	.87	.48	.75	.79	**51-52**	150	762
Public Affairs (3NXXX)	IRAQ/AFG	LO															**4**	33	33
		HI	.47	.78	.63	.53	.59	.41	.59	.41	.57	.47	.78	.31	.47	.53	**30-32**	78	78
	OTHER	LO															**3**	20	20
		HI	.54	.63	.58	.50	.54	.46	.63	.35	.54	.42	.58	.25	.38	.35	**23-24**	77	77
Security Forces (Military Police) (3PXXX)	IRAQ/AFG	LO	.83	.92	.92	.83	.92	.75	.83	.83	.83	.92	1.00	.75	.92	1.00	**12**	150	3,442
		HI	.63	.76	.68	.71	.73	.54	.73	.66	.68	.70	.80	.37	.75	.73	**40-41**	150	2,206
	OTHER	LO	.81	.88	.81	.88	.81	.63	.88	.81	.87	.88	.88	.56	.81	.81	**15-16**	150	2,508
		HI	.78	.88	.73	.70	.80	.58	.83	.60	.78	.73	.83	.49	.68	.78	**40-41**	150	1,633
Mission Support (3SXXX)	IRAQ/AFG	LO	.78	.75	.56	.56	.78	.44	.78	.56	.75	.56	.67	.33	.56	.56	**8-9**	92	92
		HI	.46	.69	.54	.51	.60	.37	.67	.43	.59	.45	.75	.29	.47	.50	**50-52**	150	345
	OTHER	LO	.79	.85	.71	.79	.86	.71	.86	.79	.86	.79	1.00	.57	.79	.79	**13-14**	121	121
		HI	.69	.77	.65	.62	.73	.56	.75	.60	.71	.62	.81	.52	.62	.62	**51-52**	150	467
Visual Information (3VXXX)	IRAQ/AFG	LO															**4**	54	54
		HI	.61	.71	.57	.57	.57	.50	.68	.50	.61	.50	.71	.43	.57	.57	**28**	81	81
	OTHER	LO	.20	.20	.40	.40	.20	.20	.20	.20	.20	.20	.40	.20	.20	.20	**5**	35	35
		HI	.69	.81	.63	.69	.69	.38	.75	.56	.81	.56	.94	.38	.56	.56	**16**	60	60
Medical (4AXXX)	IRAQ/AFG	LO	.62	.67	.62	.57	.55	.43	.71	.57	.62	.62	.71	.43	.50	.67	**20-21**	150	157
		HI	.49	.66	.68	.45	.64	.26	.62	.38	.60	.42	.77	.21	.38	.49	**52-53**	150	189
	OTHER	LO	.55	.80	.74	.74	.75	.40	.70	.50	.65	.65	.90	.35	.65	.60	**19-20**	150	155
		HI	.59	.76	.65	.57	.65	.43	.65	.54	.68	.49	.73	.41	.51	.51	**37**	150	250

RAND *MG811-T.E.6c*

Table E.6—continued

2-Digit AFSC	Deployed Location	Grade	Gather/Interpret Info	Respect Cultural Diffs	Apply Social Etiquette	Establish Credibility	Change Behavior	Verbal/Nonverbal Comm	Manage Stress	Negotiate with Others	Apply Regional Knowledge	Resolve Conflict	Self-Initiated Learning	Foreign Lang Skills	Influence Others	Establish Authority	# Respondents	Invited Sample	Population
Bioenvironmental Engineering (4BXXX)	IRAQ/AFG	LO															2	8	12
		HI	.86	.71	.71	.71	.71	.57	.86	.71	.71	.71	.86	.43	.71	.71	7	14	25
	OTHER	LO															0	16	25
		HI															3	19	39
Mental Health Services (4CXXX)	IRAQ/AFG	LO															0-3	15	21
		HI	.71	.71	.71	.57	.71	.43	.71	.43	.71	.43	1.00	.29	.57	.43	7	22	40
	OTHER	LO															2	22	30
		HI	.67	.67	.67	.67	.50	.67	.67	.50	.67	.50	.67	.33	.67	.50	6	12	23
Diet Therapy (4DXXX)	IRAQ/AFG	LO															0	13	22
		HI															1	8	13
	OTHER	LO															0	5	9
		HI															0	1	1
Public Health (4EXXX)	IRAQ/AFG	LO															0-1	8	14
		HI															2	10	21
	OTHER	LO															3	17	23
		HI	.50	.67	.67	.33	.33	.33	.50	.33	.50	.33	.50	.33	.33	.33	6	17	47
Cardiopulmonary Laboratory (4HXXX)	IRAQ/AFG	LO															3	29	39
		HI	.75	1.00	1.00	.50	1.00	.75	1.00	.50	1.00	.50	1.00	.75	.75	.75	4	8	27
	OTHER	LO															2	13	18
		HI	.50	.63	.50	.50	.50	.38	.50	.50	.63	.50	.63	.38	.63	.63	8	10	27
Physical Medicine (4JXXX)	IRAQ/AFG	LO															0	4	4
		HI															2	10	24
	OTHER	LO															0	0	0
		HI															0	3	4
Aerospace Physiology (4MXXX)	IRAQ/AFG	LO															0	2	5
		HI															2	2	11
	OTHER	LO															2-3	16	20
		HI	.83	1.00	.67	.67	.67	.67	.83	.67	.67	.67	.83	.67	.67	.67	6	13	25
Medical Service (4NXXX)	IRAQ/AFG	LO	.67	.87	.80	.73	.73	.67	.80	.73	.73	.73	.73	.60	.73	.73	15	150	271
		HI	.66	.83	.72	.63	.68	.50	.85	.60	.78	.64	.79	.47	.59	.64	45-47	150	468
	OTHER	LO	.67	.83	.61	.56	.72	.56	.72	.50	.71	.56	.83	.50	.61	.61	17-18	150	274
		HI	.51	.57	.49	.47	.53	.40	.55	.43	.58	.45	.70	.38	.43	.45	52-53	150	532
Pharmacy (4PXXX)	IRAQ/AFG	LO															2	14	20
		HI															2	14	32
	OTHER	LO															1	6	9
		HI	.63	.75	.63	.75	.63	.50	.63	.63	.63	.63	.75	.50	.63	.63	8	23	36

RAND *MG811-T.E.6d*

Table E.6—continued

2-Digit AFSC	Deployed Location	Grade	Gather/Interpret Info	Respect Cultural Diffs	Apply Social Etiquette	Establish Credibility	Change Behavior	Verbal/Nonverbal Comm	Manage Stress	Negotiate with Others	Apply Regional Knowledge	Resolve Conflict	Self-Initiated Learning	Foreign Lang Skills	Influence Others	Establish Authority	# Respondents	Invited Sample	Population
Diagnostic Imaging (4RXXX)	IRAQ/AFG	LO															**3**	21	30
		HI	1.00	1.00	1.00	1.00	1.00	1.00	1.00	1.00	1.00	1.00	1.00	1.00	1.00	1.00	**6**	17	28
	OTHER	LO															**0**	9	14
		HI															**2**	11	30
Medical Laboratory (4TXXX)	IRAQ/AFG	LO															**2**	14	26
		HI	.56	.67	.56	.44	.67	.44	.67	.44	.67	.44	.67	.33	.44	.44	**9**	23	53
	OTHER	LO															**3**	17	27
		HI	.50	.67	.67	.33	.67	.33	.67	.33	.67	.33	.67	.33	.33	.33	**6**	14	32
Optometry (4VXXX)	IRAQ/AFG	LO															**0**	2	2
		HI															**0**	3	5
	OTHER	LO															**0-1**	3	8
		HI															**0**	3	8
Dental (4YXXX)	IRAQ/AFG	LO															**2**	20	30
		HI															**2-3**	19	36
	OTHER	LO															**2**	26	38
		HI															**4-5**	24	37
Paralegal (5JXXX)	IRAQ/AFG	LO															**0**	1	1
		HI	.37	.74	.70	.48	.56	.33	.67	.33	.59	.44	.74	.30	.37	.44	**27**	67	67
	OTHER	LO															**0-1**	6	6
		HI	.50	.83	.83	.58	.75	.67	.67	.50	.58	.50	.67	.33	.50	.50	**12**	44	44
Chaplain Assistant (5RXXX)	IRAQ/AFG	LO															**3**	24	24
		HI	.57	.79	.71	.57	.64	.50	.79	.50	.64	.50	.79	.29	.43	.50	**14**	48	48
	OTHER	LO															**4**	21	21
		HI	.79	.93	.93	.71	.93	.79	.86	.57	.93	.64	1.00	.50	.57	.57	**14**	33	33
Contracting (6CXXX)	IRAQ/AFG	LO	.90	.80	.80	.78	.80	.60	.80	.80	.90	.80	1.00	.50	.70	.80	**9-10**	51	51
		HI	.64	.76	.76	.72	.67	.50	.69	.78	.72	.59	.86	.38	.62	.66	**57-58**	150	211
	OTHER	LO	.71	1.00	.86	.71	.57	.57	.57	.86	.71	.71	.86	.57	.57	.57	**7**	59	59
		HI	.63	.76	.67	.63	.61	.52	.56	.73	.66	.59	.71	.31	.55	.57	**50-51**	150	168
Financial (6FXXX)	IRAQ/AFG	LO	.60	.60	.60	.60	.60	.60	.60	.60	.60	.60	.80	.60	.60	.60	**5**	55	55
		HI	.67	.79	.54	.58	.65	.50	.63	.67	.71	.58	.71	.50	.63	.65	**23-24**	118	118
	OTHER	LO															**4**	103	103
		HI	.63	.76	.72	.67	.67	.58	.67	.65	.72	.63	.84	.47	.63	.67	**42-43**	150	156
Special Investigations (OSI) (7SXXX)	IRAQ/AFG	LO															**0**	0	0
		HI	.84	.80	.84	.80	.84	.60	.80	.76	.84	.76	.84	.48	.76	.76	**25**	121	121
	OTHER	LO															**0**	0	0
		HI	.75	.88	.88	.88	.88	.63	.88	.88	.88	.75	.88	.63	.88	.75	**8**	81	81

RAND *MG811-T.E.6e*

Table E.6—continued

2-Digit AFSC	Deployed Location	Grade	Gather/Interpret Info	Respect Cultural Diffs	Apply Social Etiquette	Establish Credibility	Change Behavior	Verbal/Nonverbal Comm	Manage Stress	Negotiate with Others	Apply Regional Knowledge	Resolve Conflict	Self-Initiated Learning	Foreign Lang Skills	Influence Others	Establish Authority	# Respondents	Invited Sample	Population
Enlisted Aide (8AXXX)	IRAQ/AFG	LO															0	0	0
		HI															0	1	1
	OTHER	LO															0	0	0
		HI															0	9	9
Military Training (8BXXX)	IRAQ/AFG	LO															0	1	1
		HI															4	16	16
	OTHER	LO															0	0	0
		HI															4	15	15
Family Support Center (8CXXX)	IRAQ/AFG	LO															0	0	0
		HI															0	0	0
	OTHER	LO															0	0	0
		HI															2	6	6
Linguist Debriefer (8DXXX)	IRAQ/AFG	LO															0	0	0
		HI															0	11	11
	OTHER	LO															0	0	0
		HI															0	0	0
First Sergeant (8FXXX)	IRAQ/AFG	LO															0	0	0
		HI	.59	.69	.67	.56	.68	.46	.67	.51	.72	.52	.76	.43	.52	.52	53-54	141	141
	OTHER	LO															0	0	0
		HI	.62	.72	.69	.57	.70	.46	.58	.49	.66	.51	.72	.42	.53	.55	47-48	150	189
USAF Honor Guard (8GXXX)	IRAQ/AFG	LO															0	0	0
		HI															0	1	1
	OTHER	LO															0	0	0
		HI															0	0	0
Postal Specialist (8MXXX)	IRAQ/AFG	LO															0	22	22
		HI	.60	.90	.70	.70	.70	.50	.80	.60	.80	.70	1.00	.50	.70	.70	10	52	52
	OTHER	LO	.71	.64	.60	.71	.60	.47	.73	.57	.79	.43	.57	.33	.50	.50	14-15	89	89
		HI	.58	.68	.58	.02	.50	.47	.67	.55	.66	.54	.74	.45	.55	.53	37-38	100	100
Recruiter (8RXXX)	IRAQ/AFG	LO															0	3	3
		HI															0-1	6	6
	OTHER	LO															0	4	4
		HI															0-3	15	15
Professional Military Education Instructor (8TXXX)	IRAQ/AFG	LO															1	0	0
		HI															0	1	1
	OTHER	LO															0	0	0
		HI															0	3	3

RAND *MG811-T.E.6f*

Table E.6—continued

2-Digit AFSC	Deployed Location	Grade	Gather/Interpret Info	Respect Cultural Diffs	Apply Social Etiquette	Establish Credibility	Change Behavior	Verbal/Nonverbal Comm	Manage Stress	Negotiate with Others	Apply Regional Knowledge	Resolve Conflict	Self-Initiated Learning	Foreign Lang Skills	Influence Others	Establish Authority	# Respondents	Invited Sample	Population
Awaiting Retraining (9AXXX)	IRAQ/AFG	LO															**1**	1	1
		HI															**0**	0	0
	OTHER	LO															**0**	4	4
		HI															**0**	0	0
Dorm Manager (9DXXX)	IRAQ/AFG	LO															**0**	0	0
		HI															**0**	9	9
	OTHER	LO															**0**	1	1
		HI															**0**	10	10
Senior Enlisted Advisor (9EXXX)	IRAQ/AFG	LO															**0**	0	0
		HI															**0**	1	1
	OTHER	LO															**0**	0	0
		HI															**3**	4	4
Group Superintendent (9GXXX)	IRAQ/AFG	LO															**0**	0	0
		HI	.70	.80	.70	.60	.70	.60	.80	.60	.80	.60	.80	.60	.60	.60	**10**	25	25
	OTHER	LO															**0**	0	0
		HI	.70	.70	.60	.50	.70	.50	.70	.50	.70	.50	.80	.40	.50	.50	**10**	23	23
Interpreter / Translator (9LXXX)	IRAQ/AFG	LO															**0**	0	0
		HI															**0**	0	0
	OTHER	LO															**0**	0	0
		HI															**0**	3	3
Technical Applications Specialist (9SXXX)	IRAQ/AFG	LO															**0**	0	0
		HI															**0-1**	0	0
	OTHER	LO															**0**	5	5
		HI	.40	.40	.20	.20	.40	.20	.40	.20	.40	.20	.40	.20	.20	.20	**5**	18	18
Basic Enlisted Trainee (9TXXX)	IRAQ/AFG	LO															**0**	13	13
		HI															**0**	0	0
	OTHER	LO															**0-1**	8	8
		HI															**0-1**	0	0
Unallotted Airman (9UXXX)	IRAQ/AFG	LO	.71	.71	.57	.50	.64	.36	.79	.50	.71	.57	.64	.29	.50	.50	**14**	0	0
		HI	.69	.56	.69	.63	.75	.50	.75	.50	.75	.63	.67	.31	.63	.69	**15-16**	0	0
	OTHER	LO	.71	.64	.71	.64	.71	.50	.71	.64	.71	.62	.86	.43	.64	.64	**13-14**	0	0
		HI	.78	.89	.56	.78	.89	.44	.78	.67	.67	.67	1.00	.33	.67	.67	**9**	0	0
TCN ESCORT	IRAQ/AFG	LO	.60	.80	.72	.68	.72	.40	.60	.64	.68	.64	.75	.40	.72	.72	**24-25**	0	0
		HI	.65	.83	.74	.57	.70	.39	.78	.48	.70	.57	.86	.39	.48	.65	**22-23**	0	0
	OTHER	LO	.77	.86	.73	.82	.70	.43	.75	.77	.73	.77	.82	.36	.77	.84	**44**	0	0
		HI	.56	.74	.67	.59	.63	.41	.63	.56	.73	.52	.74	.41	.58	.59	**26-27**	0	0

RAND *MG811-T.E.6g*

Table E.7
Officer Over-Career Training, Helpfulness Ratings, by AFSC, Grade, and Deployment Subgroups

Light Blue = Mild Helpfulness Ratings of 3.1–3.4 | Medium Blue = Medium Helpfulness Ratings of 3.5–3.9 | Dark Blue = High Helpfulness Ratings of 4.0–5.0

2-Digit AFSC	Deployed Location	Grade	Gather/Interpret Info	Respect Cultural Diffs	Apply Social Etiquette	Establish Credibility	Change Behavior	Verbal/Nonverbal Comm	Manage Stress	Negotiate with Others	Apply Regional Knowledge	Resolve Conflict	Self-Initiated Learning	Foreign Lang Skills	Influence Others	Establish Authority	# Respondents	Invited Sample	Population
All Deployed Personnel	All Deployed Personnel	All Deployed Personnel	3.1	3.4	3.2	3.0	3.2	2.7	3.2	2.8	3.2	2.7	3.3	2.2	2.7	2.7	2,344-4,573	23,020	88,304
Commander (10XX)	IRAQ/AFG	LO															0	0	0
		HI															2-4	18	18
	OTHER	LO															0	0	0
		HI	3.9	4.1	4.0	4.2	3.9	3.4	4.0	3.7	4.0	3.7	4.1	2.8	3.7	3.8	5-8	32	32
Pilot (11XX)	IRAQ/AFG	LO	2.9	2.9	2.8	2.4	2.9	2.2	3.1	1.9	3.1	1.9	3.0	1.8	2.0	1.8	11-25	150	700
		HI	2.7	2.9	3.0	2.6	2.8	1.8	2.8	2.3	3.0	2.4	3.3	1.7	2.4	2.2	29-45	150	392
	OTHER	LO	3.1	3.2	3.3	2.7	3.2	2.4	2.9	2.4	3.2	2.3	3.3	2.1	2.4	2.4	17-27	150	2,711
		HI	3.3	3.7	3.5	3.0	3.5	2.5	3.3	2.6	3.6	2.7	3.5	1.8	2.7	2.6	24-45	150	962
Navigator (12XX)	IRAQ/AFG	LO	3.0	3.0	2.9	2.8	2.0	1.9	2.9	2.5	3.1	2.5	3.0	1.6	2.6	2.5	18-28	150	254
		HI	3.0	3.3	3.2	2.9	2.9	2.6	3.3	2.6	3.3	2.7	3.5	2.1	2.7	2.3	8-25	150	161
	OTHER	LO	3.0	3.2	2.9	3.2	3.2	2.7	2.9	2.8	3.2	2.7	3.2	2.5	2.8	2.4	10-22	150	900
		HI	2.8	3.4	3.2	2.5	3.1	2.5	3.0	2.3	3.1	2.3	3.1	1.9	2.3	2.3	35-57	150	372
Space, Missile, and Command and Control (13XX)	IRAQ/AFG	LO	2.9	3.1	2.9	2.5	3.0	1.9	3.3	2.5	3.2	2.7	3.5	1.8	2.5	2.4	12-25	150	153
		HI	2.8	3.4	2.7	2.7	2.7	2.2	2.8	2.5	3.1	2.3	3.5	1.3	2.5	2.8	7-19	70	70
	OTHER	LO	3.3	3.6	3.3	3.0	3.3	2.8	3.2	2.9	3.3	2.9	3.6	2.8	3.0	2.6	12-21	150	413
		HI	3.2	3.7	3.4	3.1	3.4	2.5	3.4	2.6	3.5	2.3	3.6	2.1	2.5	2.1	16-49	150	184
Intelligence (14XX)	IRAQ/AFG	LO	3.5	3.4	3.2	2.8	3.1	2.6	3.4	2.3	3.8	2.2	3.9	2.1	2.5	2.4	14-43	150	236
		HI	3.8	3.6	3.4	2.9	3.6	2.8	3.6	2.6	4.0	2.6	3.9	1.3	2.8	2.8	7-24	68	68
	OTHER	LO	3.5	3.8	3.8	3.1	3.5	2.5	3.5	2.5	3.6	2.6	3.7	2.1	2.6	2.4	15-35	150	427
		HI	3.5	3.6	3.5	2.7	3.6	2.7	3.1	2.3	3.7	2.3	3.6	2.0	2.3	2.2	15-37	108	108
Weather (15XX)	IRAQ/AFG	LO															2-5	37	37
		HI															4-6	15	15
	OTHER	LO															2-10	38	38
		HI															0-3	9	9
International Affairs/FAO (16XX)	IRAQ/AFG	LO															4-10	6	6
		HI	3.1	3.7	3.0	3.1	3.0	2.6	3.3	2.5	3.1	2.7	3.6	1.6	2.9	2.6	14-37	39	39
	OTHER	LO															3-4	5	5
		HI	3.6	3.5	3.3	3.6	3.6	2.8	3.6	2.5	3.6	2.6	3.4	2.2	2.4	2.0	5-12	43	43

RAND *MG811-T.E.7a*

Table E.7—continued

2-Digit AFSC	Deployed Location	Grade	Gather/Interpret Info	Respect Cultural Diffs	Apply Social Etiquette	Establish Credibility	Change Behavior	Verbal/Nonverbal Comm	Manage Stress	Negotiate with Others	Apply Regional Knowledge	Resolve Conflict	Self-Initiated Learning	Foreign Lang Skills	Influence Others	Establish Authority	# Respondents	Invited Sample	Population
Logistics Commander (20XX)	IRAQ/AFG	LO															0	0	0
		HI															2-4	18	18
	OTHER	LO															0	0	0
		HI															2-5	10	10
Aircraft Maintenance (21XX)	IRAQ/AFG	LO	3.1	3.3	2.8	3.1	3.2	2.7	3.3	2.8	3.3	2.7	3.2	1.9	2.8	2.7	12-20	150	295
		HI	3.1	3.5	3.3	2.8	3.2	2.2	3.2	2.6	3.1	2.5	3.6	1.7	2.5	2.4	21-43	133	133
	OTHER	LO	3.2	3.1	3.3	2.6	3.0	2.7	3.2	2.5	3.1	2.3	3.3	2.2	2.4	2.3	13-25	150	470
		HI	3.2	3.6	3.2	2.8	2.9	2.5	3.1	2.6	3.3	2.5	3.7	2.0	2.5	2.4	22-37	150	153
Support Commander (30XX)	IRAQ/AFG	LO															0	0	0
		HI															2-9	26	26
	OTHER	LO															0	0	0
		HI	3.3	3.9	3.7	3.4	3.8	2.5	3.7	2.9	3.6	2.9	4.1	2.2	2.9	2.9	6-9	26	26
Security Forces (31XX)	IRAQ/AFG	LO	3.5	3.5	3.0	3.2	3.1	2.3	3.0	2.9	3.2	3.1	3.6	1.8	3.2	3.1	14-24	150	178
		HI															3-12	45	45
	OTHER	LO															3-9	50	50
		HI															1-2	12	12
Civil Engineer (32XX)	IRAQ/AFG	LO	3.2	3.4	3.2	3.1	3.1	2.5	3.2	2.9	3.2	2.7	3.7	1.7	2.8	2.7	13-25	150	218
		HI	2.8	3.4	3.3	3.0	3.1	2.2	3.1	2.9	3.2	2.6	3.3	1.5	2.6	2.2	18-38	98	98
	OTHER	LO	3.2	3.4	3.2	3.1	3.1	3.2	3.4	2.9	3.3	2.8	3.5	1.6	3.0	2.8	8-22	139	139
		HI	3.4	3.7	3.6	3.4	3.5	3.6	3.9	3.5	3.1	3.0	4.0	2.3	3.3	2.5	7-14	48	48
Comm-Information Systems (33XX)	IRAQ/AFG	LO	3.1	3.3	3.0	2.6	3.0	1.9	3.1	2.3	3.2	2.3	3.6	1.5	2.3	2.4	16-30	150	317
		HI	3.0	3.3	3.0	2.6	3.1	2.0	3.2	2.1	3.1	2.5	3.3	1.5	2.4	2.1	16-31	91	91
	OTHER	LO	3.4	3.4	3.5	2.7	3.4	2.1	3.2	2.5	3.2	2.8	3.5	1.8	2.5	2.6	6-18	150	397
		HI	3.3	3.5	3.3	3.0	3.5	2.7	3.5	2.6	3.4	2.6	3.6	1.9	2.7	2.5	18-38	94	94
Services (34XX)	IRAQ/AFG	LO	2.8	3.5	2.9	3.1	2.9	2.3	3.0	2.8	2.9	2.6	3.4	1.7	2.7	2.7	6-14	45	45
		HI															1-2	15	15
	OTHER	LO	3.5	3.8	3.8	3.5	3.5	3.0	3.6	3.0	3.6	2.9	3.8	2.8	3.0	2.5	6-14	87	87
		HI	3.3	3.4	3.6	3.2	3.1	2.2	3.5	2.5	3.3	2.7	3.7	2.0	3.5	2.8	5-7	18	18
Public Affairs (35XX)	IRAQ/AFG	LO															4-6	32	32
		HI															1-6	11	11
	OTHER	LO	2.9	3.0	2.8	3.1	2.8	2.5	2.8	2.7	3.3	2.9	3.9	2.0	2.9	2.6	6-9	40	40
		HI															1-3	25	25
Personnel (36XX)	IRAQ/AFG	LO															0	26	26
		HI															0	15	15
	OTHER	LO															0	41	41
		HI															0	9	9

RAND *MG811-T.E.7b*

Table E.7—continued

2-Digit AFSC	Deployed Location	Grade	Gather/Interpret Info	Respect Cultural Diffs	Apply Social Etiquette	Establish Credibility	Change Behavior	Verbal/Nonverbal Comm	Manage Stress	Negotiate with Others	Apply Regional Knowledge	Resolve Conflict	Self-Initiated Learning	Foreign Lang Skills	Influence Others	Establish Authority	# Respondents	Invited Sample	Population
Manpower-Personnel (37XX)	IRAQ/AFG	LO	2.6	3.1	2.5	2.0	2.8	1.7	3.2	1.8	2.5	1.8	3.4	1.5	1.8	2.0	6-14	51	51
		HI															3-14	36	36
	OTHER	LO	3.2	3.5	3.1	2.9	3.2	2.2	3.2	2.4	3.2	2.5	3.5	1.8	2.5	2.4	16-34	83	83
		HI	3.3	3.7	3.6	3.1	3.6	3.1	3.6	2.9	3.3	2.7	3.5	2.2	2.7	2.7	5-11	31	31
Manpower (38XX)	IRAQ/AFG	LO															0	3	3
		HI															0	1	1
	OTHER	LO															0	13	13
		HI															0	3	3
Medical Commander (40XX)	IRAQ/AFG	LO															0	0	0
		HI															2-5	7	7
	OTHER	LO															0	0	0
		HI															2-5	14	14
Health Services Administrator (41XX)	IRAQ/AFG	LO															2-4	40	40
		HI	2.7	2.8	2.8	2.4	2.5	2.0	2.6	2.3	2.6	2.5	3.2	1.6	2.3	2.3	10-14	44	44
	OTHER	LO															3-6	44	44
		HI															3-9	37	37
Biomedical Clinician (42XX)	IRAQ/AFG	LO	3.0	3.1	3.1	2.7	2.7	1.7	3.1	2.4	3.0	2.5	3.2	1.5	2.7	2.4	12-19	70	70
		HI															3-8	34	34
	OTHER	LO															2-3	48	48
		HI															1-2	17	17
Biomedical Specialists (43XX)	IRAQ/AFG	LO															3-4	32	32
		HI	3.2	3.3	3.3	2.8	3.1	2.4	3.3	2.5	3.2	2.5	3.5	1.8	2.8	2.6	6-15	49	49
	OTHER	LO															3-11	34	34
		HI															4-17	42	42
Physician (44XX)	IRAQ/AFG	LO															1-3	46	46
		HI	3.0	3.4	3.1	3.1	3.1	2.3	3.2	2.7	3.3	2.5	3.5	1.7	2.7	2.6	11-22	120	120
	OTHER	LO															0-3	48	48
		HI	3.0	3.1	2.8	2.3	2.9	2.2	2.7	2.2	2.9	2.2	3.5	1.4	2.0	1.9	8-17	77	77
Surgery (45XX)	IRAQ/AFG	LO															1	16	16
		HI	3.0	3.5	3.1	3.3	2.8	2.9	2.9	2.6	2.8	2.8	3.3	2.3	2.9	2.6	6-15	113	113
	OTHER	LO															0	4	4
		HI	3.3	3.7	3.3	3.4	3.4	2.2	3.3	2.6	3.3	2.4	3.7	1.6	2.6	2.3	5-7	38	38
Nurse (46XX)	IRAQ/AFG	LO	3.0	3.4	3.3	3.1	3.0	2.3	3.0	2.3	3.3	2.5	3.3	2.0	2.1	2.2	11-33	150	258
		HI	3.1	3.4	2.9	2.6	2.8	2.0	3.1	2.1	2.9	2.0	3.5	1.6	2.2	2.1	23-37	99	99
	OTHER	LO	3.2	3.6	3.2	3.0	3.2	2.8	3.2	2.9	3.4	2.8	3.4	2.4	2.9	2.9	23-33	150	269
		HI	3.1	3.5	3.2	2.9	3.1	2.3	3.2	2.1	3.1	2.1	3.5	1.8	2.2	2.2	17-30	94	94

RAND *MG811-T.E.7c*

Table E.7—continued

2-Digit AFSC	Deployed Location	Grade	Gather/Interpret Info	Respect Cultural Diffs	Apply Social Etiquette	Establish Credibility	Change Behavior	Verbal/Nonverbal Comm	Manage Stress	Negotiate with Others	Apply Regional Knowledge	Resolve Conflict	Self-Initiated Learning	Foreign Lang Skills	Influence Others	Establish Authority	# Respondents	Invited Sample	Population
Dental (47XX)	IRAQ/AFG	LO															0	4	4
		HI															0-3	12	12
	OTHER	LO															0-1	8	8
		HI															4-8	24	24
Aerospace Medicine (48XX)	IRAQ/AFG	LO															0-3	26	26
		HI	3.4	3.5	3.9	3.4	3.7	2.4	3.6	3.0	3.7	3.1	3.8	1.5	3.3	3.1	6-14	51	51
	OTHER	LO															2-5	57	57
		HI	3.0	3.4	3.3	3.0	3.3	2.3	3.2	2.4	3.1	2.4	3.6	1.5	2.3	2.1	14-23	81	81
Law (51XX)	IRAQ/AFG	LO															3-8	56	56
		HI	2.7	3.3	3.3	2.5	2.6	1.9	2.8	2.4	2.8	2.3	3.5	1.6	2.5	2.1	10-16	43	43
	OTHER	LO															1-3	37	37
		HI	2.7	3.5	2.6	3.1	2.9	2.0	2.7	2.9	2.9	3.0	3.5	1.6	2.9	2.8	9-14	49	49
Chaplain (52XX)	IRAQ/AFG	LO	3.2	3.6	3.3	3.0	3.4	2.3	3.0	2.4	3.6	2.6	3.6	1.3	2.8	2.3	6-12	43	43
		HI															4-17	49	49
	OTHER	LO															3-9	43	43
		HI															3-9	25	25
Program Director (60XX)	IRAQ/AFG	LO															0	0	0
		HI															0	1	1
	OTHER	LO															0	0	0
		HI															0	0	0
Scientific / Research (61XX)	IRAQ/AFG	LO															0-2	9	9
		HI															1-3	4	4
	OTHER	LO	2.4	2.2	2.4	2.0	2.6	2.0	2.8	1.6	2.2	1.6	2.6	2.0	1.6	2.0	5	15	15
		HI															2-4	16	16
Developmental Engineering (62XX)	IRAQ/AFG	LO															1-4	21	21
		HI															0-1	10	10
	OTHER	LO															3-6	31	31
		HI															0	5	5
Acquisition (63XX)	IRAQ/AFG	LO															4-8	33	33
		HI															4-6	42	42
	OTHER	LO															2	31	31
		HI															3	17	17
Contracting (64XX)	IRAQ/AFG	LO	3.4	3.8	3.3	3.4	3.1	2.6	3.3	3.6	3.5	3.1	4.0	1.6	2.9	2.9	8-27	111	111
		HI	3.1	3.5	3.3	3.3	3.3	2.8	3.2	3.6	3.3	2.8	3.2	1.7	3.0	3.0	7-18	52	52
	OTHER	LO	3.5	3.7	3.7	3.7	3.4	3.7	3.5	3.9	3.4	3.1	3.9	2.6	3.4	3.1	7-20	69	69
		HI															1-3	4	4

RAND *MG811-T.E.7d*

Table E.7—continued

2-Digit AFSC	Deployed Location	Grade	Gather/Interpret Info	Respect Cultural Diffs	Apply Social Etiquette	Establish Credibility	Change Behavior	Verbal/Nonverbal Comm	Manage Stress	Negotiate with Others	Apply Regional Knowledge	Resolve Conflict	Self-Initiated Learning	Foreign Lang Skills	Influence Others	Establish Authority	# Respondents	Invited Sample	Population
Finance (65XX)	IRAQ/AFG	LO															2-7	43	43
		HI															1-3	37	37
	OTHER	LO															1-4	22	22
		HI															4-7	17	17
Special Investigations (71XX)	IRAQ/AFG	LO															4	30	30
		HI															1-3	14	14
	OTHER	LO															3-4	17	17
		HI															0-1	4	4
Commander—USAFA Cadet Squadron (80XX)	IRAQ/AFG	LO															0	0	0
		HI															0	0	0
	OTHER	LO															0	0	0
		HI															0	1	1
Instructor (81XX)	IRAQ/AFG	LO															1	4	4
		HI															0-3	8	8
	OTHER	LO															0	6	6
		HI															0	8	8
Academic Program Manager (82XX)	IRAQ/AFG	LO															0	1	1
		HI															0	3	3
	OTHER	LO															0	1	1
		HI															0	3	3
Recruiting Service (83XX)	IRAQ/AFG	LO															0	4	4
		HI															0	0	0
	OTHER	LO															0	2	2
		HI															0	0	0
Historian (84XX)	IRAQ/AFG	LO															0	0	0
		HI															0	0	0
	OTHER	LO															0	0	0
		HI															0	0	0
USAF Honor Guard (85XX)	IRAQ/AFG	LO															0	0	0
		HI															0	0	0
	OTHER	LO															0	0	0
		HI															0	1	1
Command and Control (86XX)	IRAQ/AFG	LO															0	2	2
		HI															0-2	3	3
	OTHER	LO															1-2	3	3
		HI															0-1	5	5

RAND *MG811-T.E.7e*

Table E.7—continued

2-Digit AFSC	Deployed Location	Grade	Gather/Interpret Info	Respect Cultural Diffs	Apply Social Etiquette	Establish Credibility	Change Behavior	Verbal/Nonverbal Comm	Manage Stress	Negotiate with Others	Apply Regional Knowledge	Resolve Conflict	Self-Initiated Learning	Foreign Lang Skills	Influence Others	Establish Authority	# Respondents	Invited Sample	Population
Inspector General (87XX)	IRAQ/AFG	LO															0	0	0
		HI															0	0	0
	OTHER	LO															0	0	0
		HI															0	1	1
Protocol / Aide (88XX)	IRAQ/AFG	LO															0-2	1	1
		HI															0-1	0	0
	OTHER	LO															2-3	0	0
		HI															0	0	0
General Officer (90XX)	IRAQ/AFG	LO															0	0	0
		HI															2-3	10	10
	OTHER	LO															0	0	0
		HI															0	11	11
Commander (91XX)	IRAQ/AFG	LO															0	0	0
		HI															1-4	13	13
	OTHER	LO															0	0	0
		HI															4-7	18	18
Student (92XX)	IRAQ/AFG	LO															0-2	22	22
		HI															0	7	7
	OTHER	LO															0	65	65
		HI															0	12	12
Patient (93XX)	IRAQ/AFG	LO															0	0	0
		HI															0	0	0
	OTHER	LO															0	0	0
		HI															0	0	0
Nuclear Weapons Custodian (94XX)	IRAQ/AFG	LO															0	0	0
		HI															0	0	0
	OTHER	LO															0	0	0
		HI															0	0	0
Non-extended Active Duty (95XX)	IRAQ/AFG	LO															0	0	0
		HI															0	0	0
	OTHER	LO															0	0	0
		HI															0	1	1
Unclassified Officer (96XX)	IRAQ/AFG	LO															0	0	0
		HI															0	0	0
	OTHER	LO															0	2	2
		HI															0	0	0

RAND *MG811-T.E.7f*

Table E.7—continued

2-Digit AFSC	Deployed Location	Grade	Gather/Interpret Info	Respect Cultural Diffs	Apply Social Etiquette	Establish Credibility	Change Behavior	Verbal/Nonverbal Comm	Manage Stress	Negotiate with Others	Apply Regional Knowledge	Resolve Conflict	Self-Initiated Learning	Foreign Lang Skills	Influence Others	Establish Authority	# Respondents	Invited Sample	Population
Executive Officer above Wing Level (97XX)	IRAQ/AFG	LO															1	3	3
		HI															1-2	12	12
	OTHER	LO															0-1	6	6
		HI															0	7	7

RAND *MG811-T.E.7g*

Table E.8
Enlisted Over-Career Training, Helpfulness Ratings, by AFSC, Grade, and Deployment Subgroups

Light Blue = Mild Helpfulness Ratings of 3.1–3.4 | Medium Blue = Medium Helpfulness Ratings of 3.5–3.9 | Dark Blue = High Helpfulness Ratings of 4.0–5.0

2-Digit AFSC	Deployed Location	Grade	Gather/Interpret Info	Respect Cultural Diffs	Apply Social Etiquette	Establish Credibility	Change Behavior	Verbal/Nonverbal Comm	Manage Stress	Negotiate with Others	Apply Regional Knowledge	Resolve Conflict	Self-Initiated Learning	Foreign Lang Skills	Influence Others	Establish Authority	# Respondents	Invited Sample	Population
All Deployed Personnel	All Deployed Personnel	All Deployed Personnel	3.1	3.4	3.2	3.0	3.2	2.7	3.2	2.8	3.2	2.7	3.3	2.2	2.7	2.7	2,344-4,573	23,020	88,304
Aircrew Operations (1AXXX)	IRAQ/AFG	LO	3.4	3.2	3.1	3.2	3.1	2.9	3.3	3.1	3.5	3.1	3.3	2.4	3.2	3.3	7-12	150	218
		HI	3.4	3.4	3.5	3.4	3.4	2.7	3.4	3.1	3.6	3.0	3.6	1.9	3.1	2.9	10-24	150	633
	OTHER	LO	3.2	3.4	3.4	3.2	3.4	2.9	3.4	3.0	3.2	2.7	3.1	2.6	2.6	2.5	10-21	150	1,248
		HI	3.1	3.2	3.4	3.0	3.2	2.8	3.2	2.7	3.1	2.4	3.1	2.6	2.4	2.2	17-34	150	1,431
Command & Control Systems Operations (1CXXX)	IRAQ/AFG	LO	3.1	3.4	3.1	2.7	3.2	2.0	2.8	2.4	2.8	2.3	3.0	1.9	2.3	2.3	9-15	150	859
		HI	2.8	3.2	2.8	2.8	2.8	2.5	3.0	2.4	3.1	2.6	3.2	2.0	2.6	2.5	27-40	150	983
	OTHER	LO	3.1	3.2	3.3	2.9	3.2	3.0	3.3	2.8	2.9	2.7	3.1	2.5	2.8	2.7	11-18	150	554
		HI	3.2	3.5	3.3	3.0	3.3	3.0	3.3	3.0	3.4	2.8	3.7	2.3	2.7	2.8	27-50	150	676
Intelligence (1NXXX)	IRAQ/AFG	LO	2.6	3.0	2.8	2.5	2.6	2.3	3.1	2.5	3.4	2.3	3.7	2.0	2.9	2.4	6-14	150	290
		HI	3.6	3.3	3.5	2.7	3.4	3.0	3.5	2.5	3.7	2.2	3.9	2.1	2.4	2.2	14-47	150	313
	OTHER	LO	3.4	3.7	3.6	2.8	3.3	2.8	3.3	2.8	3.6	2.4	3.7	1.6	2.6	2.1	5-16	150	436
		HI	3.1	3.5	3.4	2.8	3.2	2.5	3.4	2.6	3.5	2.4	3.6	1.7	2.4	2.3	20-27	150	437
Safety (1SXXX)	IRAQ/AFG	LO															0	0	0
		HI															2-11	39	39
	OTHER	LO															0	1	1
		HI															2-10	41	41
Aircrew Protection (1TXXX)	IRAQ/AFG	LO	2.8	3.2	2.7	3.0	2.6	2.8	3.2	2.6	2.6	3.0	3.5	2.4	2.6	2.6	5-6	115	115
		HI	3.4	3.8	3.6	3.0	3.4	2.9	3.9	2.7	3.8	2.8	3.8	2.4	2.9	3.1	9-18	118	118
	OTHER	LO															4-9	150	276
		HI	3.3	3.5	3.3	3.1	3.4	2.9	3.5	2.8	3.3	2.7	3.8	2.4	2.6	2.6	17-35	150	290
Weather (1WXXX)	IRAQ/AFG	LO															3-4	68	68
		HI	2.9	3.2	2.7	2.7	2.7	2.2	2.9	2.3	3.2	2.3	3.4	1.8	2.6	2.5	18-44	150	217
	OTHER	LO															2-3	54	54
		HI	3.3	3.5	3.5	2.8	3.2	2.7	3.3	2.6	3.5	2.2	3.5	1.9	2.5	2.3	16-37	127	127
Manned Aerospace Maintenance (2AXXX)	IRAQ/AFG	LO															2-7	150	2,505
		HI	2.8	3.5	3.1	2.7	3.0	2.6	3.2	2.4	3.1	2.4	3.2	1.8	2.4	2.6	21-46	150	2,396
	OTHER	LO															4-11	150	5,727
		HI	2.9	3.2	2.9	2.6	3.0	2.4	3.0	2.5	3.0	2.4	3.2	2.2	2.4	2.5	27-39	150	5,923

RAND *MG811-T.E.8a*

Table E.8—continued

2-Digit AFSC	Deployed Location	Grade	Gather/Interpret Info	Respect Cultural Diffs	Apply Social Etiquette	Establish Credibility	Change Behavior	Verbal/Nonverbal Comm	Manage Stress	Negotiate with Others	Apply Regional Knowledge	Resolve Conflict	Self-Initiated Learning	Foreign Lang Skills	Influence Others	Establish Authority	# Respondents	Invited Sample	Population
Communications & Electronics (2EXXX)	IRAQ/AFG	LO	3.1	3.3	3.1	2.9	3.2	2.4	3.2	2.8	3.2	3.1	3.4	2.5	2.9	3.0	**11-19**	150	705
		HI	2.7	3.2	2.9	2.6	2.8	2.3	3.1	2.3	2.8	2.3	3.0	1.8	2.4	2.5	**33-58**	150	596
	OTHER	LO	3.4	3.7	3.5	3.0	3.2	3.1	3.6	2.5	3.6	3.3	3.4	2.4	2.6	2.6	**5-22**	150	654
		HI	3.2	3.5	3.6	3.1	3.5	2.4	3.5	2.6	3.2	2.5	3.4	2.0	2.3	2.4	**22-43**	150	689
Fuels (2FXXX)	IRAQ/AFG	LO	2.9	3.0	2.8	2.7	2.9	2.9	3.3	2.9	3.1	2.7	3.0	2.7	2.8	2.9	**9-19**	150	209
		HI	3.2	3.4	3.2	3.0	3.2	2.6	3.3	3.0	3.1	2.8	3.2	2.3	2.9	2.8	**24-46**	150	264
	OTHER	LO	2.9	2.9	2.8	2.7	2.9	2.9	3.1	2.7	3.0	2.6	2.7	2.5	2.5	2.5	**8-12**	150	668
		HI	3.1	3.3	3.2	3.0	3.2	2.7	3.2	2.8	3.2	2.8	3.2	2.3	2.6	2.8	**28-57**	150	709
Logistics Plans (2GXXX)	IRAQ/AFG	LO															**0-1**	22	22
		HI	2.6	3.0	2.8	2.4	2.6	2.3	2.8	2.2	2.6	2.2	2.8	2.1	2.2	2.3	**16-24**	87	87
	OTHER	LO															**3-5**	37	37
		HI	3.2	3.4	3.4	2.7	3.4	2.7	3.5	2.4	3.3	2.5	3.4	2.1	2.3	2.2	**13-31**	119	119
Missile & Space Systems Maintenance (2MXXX)	IRAQ/AFG	LO															**2-3**	59	59
		HI															**2**	22	22
	OTHER	LO															**0**	39	39
		HI															**1**	7	7
Precision Measurement Equipment Lab (2PXXX)	IRAQ/AFG	LO															**0**	5	5
		HI															**0-2**	5	5
	OTHER	LO															**0-4**	42	42
		HI	2.8	3.3	3.1	2.5	3.1	2.0	3.0	2.3	2.9	2.5	3.0	1.9	2.1	2.4	**9-20**	61	61
Maintenance Management Systems (2RXXX)	IRAQ/AFG	LO															**0-3**	42	42
		HI															**2-8**	35	35
	OTHER	LO															**2-6**	92	92
		HI	2.9	3.5	3.4	2.9	3.2	2.5	3.4	2.5	3.1	2.5	3.3	2.3	2.5	2.6	**10-24**	100	100
Supply (2SXXX)	IRAQ/AFG	LO	2.6	2.7	2.8	2.3	2.5	2.6	3.2	2.6	2.5	2.2	2.7	1.8	2.2	2.4	**6-11**	150	374
		HI	3.1	3.5	3.2	2.9	3.3	2.8	3.2	2.8	3.1	2.9	3.4	2.4	2.7	2.7	**18-32**	150	643
	OTHER	LO	3.1	3.3	3.4	3.1	3.0	3.1	3.3	3.1	3.4	3.1	3.5	2.5	3.0	3.1	**6-10**	150	650
		HI	3.1	3.6	3.4	3.2	3.1	2.7	3.5	2.8	3.2	2.8	3.3	1.9	2.6	2.7	**16-30**	150	1,067
Transportation & Vehicle Maintenance (2TXXX)	IRAQ/AFG	LO	3.3	3.2	3.3	3.1	3.3	3.1	3.3	3.1	3.2	3.1	3.3	2.8	3.1	3.2	**13-19**	150	1,232
		HI	3.1	3.1	3.0	3.0	3.0	2.4	3.2	2.7	3.2	2.8	3.4	1.9	2.7	2.7	**29-53**	150	1,219
	OTHER	LO	3.8	3.4	3.4	3.3	3.4	3.6	3.9	3.5	3.6	3.3	3.4	2.3	3.1	3.1	**6-9**	150	1,207
		HI	3.2	3.4	3.6	3.2	3.3	3.0	3.5	3.1	3.4	3.0	3.4	2.4	2.8	2.8	**23-38**	150	1,275
Munitions & Weapons (2WXXX)	IRAQ/AFG	LO	2.7	2.9	2.9	2.5	2.6	2.5	2.8	2.4	3.0	2.6	2.8	2.0	2.5	2.4	**12-18**	150	768
		HI	2.9	3.1	3.2	2.7	3.0	2.6	3.1	2.6	3.1	2.6	3.1	2.2	2.6	2.8	**19-41**	150	714
	OTHER	LO	3.8	3.7	3.6	3.6	3.3	3.2	3.6	3.4	3.6	3.2	3.8	2.9	3.3	3.3	**10-15**	150	980
		HI	2.7	3.1	3.1	2.8	2.9	2.4	3.1	2.5	2.8	2.5	2.9	2.3	2.6	2.4	**25-41**	150	905

RAND *MG811-T.E.8b*

Table E.8—continued

2-Digit AFSC	Deployed Location	Grade	Gather/Interpret Info	Respect Cultural Diffs	Apply Social Etiquette	Establish Credibility	Change Behavior	Verbal/Nonverbal Comm	Manage Stress	Negotiate with Others	Apply Regional Knowledge	Resolve Conflict	Self-Initiated Learning	Foreign Lang Skills	Influence Others	Establish Authority	# Respondents	Invited Sample	Population
Information Management (3AXXX)	IRAQ/AFG	LO	3.0	3.1	3.1	2.8	3.3	2.5	3.2	2.5	3.0	2.9	2.9	2.6	2.9	3.0	**9-14**	150	220
		HI	3.3	3.4	3.2	2.9	3.2	2.7	3.5	2.6	3.4	3.0	3.5	2.1	2.8	3.1	**18-39**	150	539
	OTHER	LO	3.2	3.3	3.5	3.3	3.2	3.0	3.3	3.1	3.1	3.2	3.1	2.6	3.0	2.8	**10-15**	150	329
		HI	3.3	3.8	3.7	3.1	3.3	3.0	3.5	2.4	3.4	2.3	3.7	1.9	2.4	2.2	**15-38**	150	756
Communications & Computer Systems (3CXXX)	IRAQ/AFG	LO	3.0	3.2	2.8	2.8	2.9	2.4	3.1	2.8	2.8	2.9	2.8	2.2	3.3	3.2	**11-18**	150	405
		HI	3.1	3.4	3.1	3.0	3.1	2.8	3.3	2.8	3.2	2.9	3.3	2.1	2.0	2.8	**12-31**	150	027
	OTHER	LO	3.6	3.3	3.3	3.0	3.1	2.9	3.3	2.7	3.6	3.0	3.3	2.1	2.6	2.9	**7-11**	150	617
		HI	2.6	3.2	3.2	2.5	3.3	2.2	3.2	2.6	2.7	2.2	3.0	1.7	2.3	2.2	**18-34**	150	1,026
Civil Engineering (3EXXX)	IRAQ/AFG	LO	3.1	2.9	2.9	2.9	3.1	2.6	3.3	2.5	2.9	3.1	3.0	2.5	2.6	2.6	**8-13**	150	1,839
		HI	3.1	3.3	3.3	3.0	3.1	2.5	3.2	2.9	3.3	2.8	3.3	2.1	2.9	3.0	**33-52**	150	1,826
	OTHER	LO	3.2	3.3	3.1	3.2	3.1	2.9	3.1	2.9	3.2	2.7	3.2	2.4	2.8	2.8	**5-13**	150	2,039
		HI	3.1	3.3	3.3	3.2	3.3	2.5	3.2	2.6	3.1	2.7	3.4	2.1	2.6	2.7	**17-35**	150	1,919
Historian (3HXXX)	IRAQ/AFG	LO															**0**	0	0
		HI															**0-1**	5	5
	OTHER	LO															**0**	0	0
		HI															**2**	7	7
Services (3MXXX)	IRAQ/AFG	LO	3.3	3.4	3.3	2.9	3.1	2.4	3.4	3.0	3.5	3.0	3.4	1.9	2.9	3.1	**7-19**	150	208
		HI	3.0	3.2	3.1	2.9	3.0	2.6	3.1	2.7	2.8	2.8	3.1	2.2	2.7	3.0	**20-34**	150	268
	OTHER	LO															**4-13**	150	703
		HI	3.4	3.7	3.5	3.5	3.4	3.1	3.5	3.2	3.3	3.0	3.5	2.5	2.9	3.0	**26-47**	150	762
Public Affairs (3NXXX)	IRAQ/AFG	LO															**2-4**	33	33
		HI	3.2	3.4	3.1	2.9	3.1	2.4	3.3	2.5	3.1	2.5	3.5	1.6	2.5	2.4	**11-26**	78	78
	OTHER	LO															**1-2**	20	20
		HI	3.8	3.8	3.4	3.4	3.8	3.2	3.8	3.5	3.6	3.4	4.0	1.8	3.0	3.0	**5-17**	77	77
Security Forces (Military Police) (3PXXX)	IRAQ/AFG	LO	3.5	3.8	3.0	3.5	3.1	2.5	3.3	3.4	3.5	3.8	3.7	1.7	3.3	3.7	**9-12**	150	3,442
		HI	3.6	3.7	3.4	3.5	3.5	2.7	3.8	3.2	3.7	3.4	3.6	2.1	3.3	3.5	**15-34**	150	2,206
	OTHER	LO	3.6	3.8	3.8	3.8	3.6	3.4	3.7	3.4	3.8	3.5	3.5	2.6	3.7	3.7	**8-16**	150	2,508
		HI	3.5	3.6	3.2	3.1	3.1	2.8	3.3	3.1	3.3	3.1	3.4	2.6	2.9	3.1	**20-37**	150	1,633
Mission Support (3SXXX)	IRAQ/AFG	LO															**4-7**	92	92
		HI	3.1	3.3	3.1	2.9	3.0	2.8	3.2	2.6	3.1	2.8	3.4	2.1	2.6	2.6	**15-36**	150	345
	OTHER	LO	3.8	3.7	3.7	3.5	3.8	3.2	3.6	3.4	3.5	3.1	3.6	2.8	2.6	2.5	**8-13**	121	121
		HI	3.3	3.6	3.4	3.3	3.2	2.8	3.4	2.9	3.4	2.7	3.6	2.0	2.9	2.7	**22-40**	150	467
Visual Information (3VXXX)	IRAQ/AFG	LO															**1-3**	54	54
		HI	3.3	3.5	3.1	3.3	3.3	2.8	3.5	2.8	3.4	2.7	3.6	1.9	2.9	2.8	**11-23**	81	81
	OTHER	LO															**0-3**	35	35
		HI	3.3	3.6	3.3	3.2	3.5	3.1	3.5	2.9	3.1		3.3	2.6	2.6	2.6	**7-12**	60	60

RAND *MG811-T.E.8c*

Table E.8—continued

2-Digit AFSC	Deployed Location	Grade	Gather/Interpret Info	Respect Cultural Diffs	Apply Social Etiquette	Establish Credibility	Change Behavior	Verbal/Nonverbal Comm	Manage Stress	Negotiate with Others	Apply Regional Knowledge	Resolve Conflict	Self-Initiated Learning	Foreign Lang Skills	Influence Others	Establish Authority	# Respondents	Invited Sample	Population
Medical (4AXXX)	IRAQ/AFG	LO	3.1	3.3	3.1	3.0	3.0	3.0	3.4	2.7	3.3	2.8	3.6	2.4	3.0	2.7	8-16	150	157
		HI	3.4	3.6	3.5	3.2	3.5	2.9	3.5	2.7	3.5	3.0	3.5	2.3	2.9	2.8	12-40	150	189
	OTHER	LO	3.4	3.7	3.2	3.0	3.3	2.9	3.5	2.9	3.2	3.1	3.4	2.3	2.8	2.7	7-15	150	155
		HI	3.4	3.5	3.6	3.0	3.3	3.2	3.6	2.6	3.3	2.6	3.5	2.4	2.5	2.6	13-28	150	250
Bioenvironmental Engineering (4BXXX)	IRAQ/AFG	LO															2	8	12
		HI															2-6	14	25
	OTHER	LO															0	16	25
		HI															2-3	19	39
Mental Health Services (4CXXX)	IRAQ/AFG	LO															1-3	15	21
		HI															2-5	22	40
	OTHER	LO															1-2	22	30
		HI															1-5	12	23
Diet Therapy (4DXXX)	IRAQ/AFG	LO															0	13	22
		HI															1	8	13
	OTHER	LO															0	5	9
		HI															0	1	1
Public Health (4EXXX)	IRAQ/AFG	LO															0	8	14
		HI															2	10	21
	OTHER	LO															2-3	17	23
		HI															2-5	17	47
Cardiopulmonary Laboratory (4HXXX)	IRAQ/AFG	LO															1-3	29	39
		HI															3-4	8	27
	OTHER	LO															1-2	13	18
		HI															4-6	10	27
Physical Medicine (4JXXX)	IRAQ/AFG	LO															0	4	4
		HI															2	10	24
	OTHER	LO															0	0	0
		HI															0	3	4
Aerospace Physiology (4MXXX)	IRAQ/AFG	LO															0	2	5
		HI															2	2	11
	OTHER	LO															2	16	20
		HI															4-5	13	25
Medical Service (4NXXX)	IRAQ/AFG	LO	2.6	3.4	2.9	2.9	3.0	2.1	3.0	2.6	2.8	2.8	3.3	1.9	2.6	2.5	8-14	150	271
		HI	2.9	3.5	3.3	3.1	3.3	2.6	3.5	2.5	3.1	2.6	3.4	2.0	2.7	2.7	18-36	150	468
	OTHER	LO	2.9	3.5	3.2	2.8	3.3	2.5	3.5	2.4	2.8	2.4	3.5	2.0	2.4	2.3	8-14	150	274
		HI	3.2	3.6	3.1	3.0	3.3	2.5	3.3	2.4	3.3	2.7	3.9	1.8	2.6	2.5	17-39	150	532

RAND *MG811-T.E.8d*

Table E.8—continued

2-Digit AFSC	Deployed Location	Grade	Gather/Interpret Info	Respect Cultural Diffs	Apply Social Etiquette	Establish Credibility	Change Behavior	Verbal/Nonverbal Comm	Manage Stress	Negotiate with Others	Apply Regional Knowledge	Resolve Conflict	Self-Initiated Learning	Foreign Lang Skills	Influence Others	Establish Authority	# Respondents	Invited Sample	Population
Pharmacy (4PXXX)	IRAQ/AFG	LO															1-2	14	20
		HI															1-2	14	32
	OTHER	LO															1	6	9
		HI	3.4	3.8	3.2	3.2	3.2	2.2	3.2	2.5	3.5	2.3	4.0	2.2	2.5	2.5	5-6	23	36
Diagnostic Imaging (4RXXX)	IRAQ/AFG	LO															3	21	30
		HI	2.2	2.7	2.7	2.7	2.8	2.2	2.8	2.2	2.8	2.7	2.5	1.8	2.7	2.2	5-6	17	28
	OTHER	LO															0	9	14
		HI															0-2	11	30
Medical Laboratory (4TXXX)	IRAQ/AFG	LO															1-2	14	26
		HI															3-7	23	53
	OTHER	LO															1-3	17	27
		HI															2-4	14	32
Optometry (4VXXX)	IRAQ/AFG	LO															0	2	2
		HI															0	3	5
	OTHER	LO															0-1	3	8
		HI															0	3	8
Dental (4YXXX)	IRAQ/AFG	LO															1-2	20	30
		HI															2	19	36
	OTHER	LO															2	26	38
		HI															3-5	24	37
Paralegal (5JXXX)	IRAQ/AFG	LO															0	1	1
		HI	3.2	3.3	3.3	3.0	3.3	2.6	3.3	2.6	2.9	2.5	3.1	1.9	2.2	2.1	8-20	67	67
	OTHER	LO															0-1	6	6
		HI	3.3	3.8	3.3	3.1	3.6	3.0	3.1	2.9	3.2	2.8	3.2	2.4	2.9	3.0	5-12	44	44
Chaplain Assistant (5RXXX)	IRAQ/AFG	LO															1-3	24	24
		HI															4-12	48	48
	OTHER	LO															2-4	21	21
		HI	3.6	3.7	3.6	3.1	3.5	3.5	3.7	3.3	3.5	2.7	3.9	2.7	2.9	2.4	6-13	33	33
Contracting (6CXXX)	IRAQ/AFG	LO															4-9	51	51
		HI	3.5	3.8	3.5	3.7	3.5	2.8	3.4	3.9	3.6	3.0	3.9	2.0	3.0	3.0	19-51	150	211
	OTHER	LO															4-7	59	59
		HI	3.6	3.8	3.8	3.8	3.6	3.5	3.7	3.9	3.7	3.4	3.6	2.3	3.2	3.3	15-44	150	168
Financial (6FXXX)	IRAQ/AFG	LO															3-4	55	55
		HI	3.1	3.5	2.9	3.1	3.0	2.3	3.4	2.8	3.2	2.3	3.7	1.8	2.5	2.2	11-19	118	118
	OTHER	LO															3-4	103	103
		HI	3.2	3.5	3.4	3.2	3.3	3.0	3.3	3.1	3.0	3.0	3.8	1.9	3.0	2.7	17-36	150	156

RAND *MG811-T.E.8e*

Table E.8—continued

2-Digit AFSC	Deployed Location	Grade	Gather/Interpret Info	Respect Cultural Diffs	Apply Social Etiquette	Establish Credibility	Change Behavior	Verbal/Nonverbal Comm	Manage Stress	Negotiate with Others	Apply Regional Knowledge	Resolve Conflic:	Self-Initiated Learning	Foreign Lang Skills	Influence Others	Establish Authority	# Respondents	Invited Sample	Population
Special Investigations (OSI) (7SXXX)	IRAQ/AFG	LO															0	0	0
		HI	3.8	3.7	3.7	3.6	3.7	3.2	3.5	3.3	3.6	3.1	3.8	2.5	3.3	3.1	12-20	121	121
	OTHER	LO															0	0	0
		HI															1-8	81	81
Enlisted Aide (8AXXX)	IRAQ/AFG	LO															0	0	0
		HI															0	1	1
	OTHER	LO															0	0	0
		HI															0	9	9
Military Training (8BXXX)	IRAQ/AFG	LO															0	1	1
		HI															2-4	16	16
	OTHER	LO															0	0	0
		HI															1-4	15	15
Family Support Center (8CXXX)	IRAQ/AFG	LO															0	0	0
		HI															0	0	0
	OTHER	LO															0	0	0
		HI															1	6	6
Linguist Debriefer (8DXXX)	IRAQ/AFG	LO															0	0	0
		HI															3-5	11	11
	OTHER	LO															0	0	0
		HI															0	0	0
First Sergeant (8FXXX)	IRAQ/AFG	LO															0	0	0
		HI	2.8	3.2	3.2	2.8	3.1	2.3	3.3	2.5	2.9	2.4	3.3	1.9	2.6	2.7	21-39	141	141
	OTHER	LO															0	0	0
		HI	3.0	3.4	3.2	3.0	3.1	2.5	3.0	2.6	2.9	2.7	3.3	2.0	2.6	2.4	20-38	150	189
USAF Honor Guard (8GXXX)	IRAQ/AFG	LO															0	0	0
		HI															0	1	1
	OTHER	LO															0	0	0
		HI															0	0	0
Postal Specialist (8MXXX)	IRAQ/AFG	LO															0	22	22
		HI															4-9	52	52
	OTHER	LO	3.2	3.4	3.3	3.4	3.1	3.1	3.1	3.1	3.3	3.3	3.5	2.7	3.2	3.1	6-11	89	89
		HI	3.1	3.3	3.2	3.1	3.1	2.8	3.2	3.0	3.2	2.7	3.5	2.2	2.9	2.8	16-30	100	100
Recruiter (8RXXX)	IRAQ/AFG	LO															0	3	3
		HI															0-1	6	6
	OTHER	LO															0	4	4
		HI															0-3	15	15

RAND *MG811-T.E.8f*

Table E.8—continued

2-Digit AFSC	Deployed Location	Grade	Gather/Interpret Info	Respect Cultural Diffs	Apply Social Etiquette	Establish Credibility	Change Behavior	Verbal/Nonverbal Comm	Manage Stress	Negotiate with Others	Apply Regional Knowledge	Resolve Conflict	Self-Initiated Learning	Foreign Lang Skills	Influence Others	Establish Authority	# Respondents	Invited Sample	Population
Professional Military Education Instructor (8TXXX)	IRAQ/AFG	LO															1	0	0
		HI															0	1	1
	OTHER	LO															0	0	0
		HI															0	3	3
Awaiting Retraining (9AXXX)	IRAQ/AFG	LO															1	1	1
		HI															0	0	0
	OTHER	LO															0	4	4
		HI															0	0	0
Dorm Manager (9DXXX)	IRAQ/AFG	LO															0	0	0
		HI															0	9	9
	OTHER	LO															0	1	1
		HI															0	10	10
Senior Enlisted Advisor (9EXXX)	IRAQ/AFG	LO															0	0	0
		HI															0	1	1
	OTHER	LO															0	0	0
		HI															1-3	4	4
Group Superintendent (9GXXX)	IRAQ/AFG	LO															0	0	0
		HI	2.8	3.3	3.3	3.0	3.4	2.6	3.1	2.4	3.1	2.6	3.6	1.8	2.3	2.3	6-9	25	25
	OTHER	LO															0	0	0
		HI															4-9	23	23
Interpreter/Translator (9LXXX)	IRAQ/AFG	LO															0	0	0
		HI															0	0	0
	OTHER	LO															0	0	0
		HI															0	3	3
Technical Applications Specialist (9SXXX)	IRAQ/AFG	LO															0	0	0
		HI															0-1	0	0
	OTHER	LO															0	5	5
		HI															1-3	18	18
Basic Enlisted Trainee (9TXXX)	IRAQ/AFG	LO															0	13	13
		HI															0	0	0
	OTHER	LO															0-1	8	8
		HI															0-1	0	0
Unallotted Airman (9UXXX)	IRAQ/AFG	LO	3.3	4.0	3.1	3.9	3.4	2.8	3.3	3.7	3.2	3.4	3.5	1.6	3.4	3.7	5-12	0	0
		HI	3.5	3.2	3.0	3.2	3.0	3.1	3.6	2.6	3.2	3.1	3.7	2.6	3.0	3.4	5-11	0	0
	OTHER	LO	3.1	3.3	3.4	2.7	3.2	2.9	3.5	2.9	3.1	2.9	3.2	1.7	2.7	3.0	6-13	0	0
		HI															3-8	0	0

RAND *MG811-T.E.8g*

Table E.8—continued

2-Digit AFSC	Deployed Location	Grade	Gather/Interpret Info	Respect Cultural Diffs	Apply Social Etiquette	Establish Credibility	Change Behavior	Verbal/Nonverbal Comm	Manage Stress	Negotiate with Others	Apply Regional Knowledge	Resolve Conflict	Self-Initiated Learning	Foreign Lang Skills	Influence Others	Establish Authority	# Respondents	Invited Sample	Population
TCN ESCORT	IRAQ/AFG	LO	3.4	3.2	3.3	3.1	3.1	2.7	3.3	3.0	3.5	3.3	3.6	2.3	3.2	3.4	7-17	0	0
		HI	3.5	3.3	2.8	3.5	3.3	2.8	3.5	3.0	3.2	3.0	3.4	2.0	2.9	4.0	8-20	0	0
	OTHER	LO	3.3	3.2	3.2	3.1	3.2	2.9	3.4	3.0	3.2	3.0	3.5	2.4	3.0	3.2	19-35	0	0
		HI	3.2	3.5	3.0	3.2	3.2		3.2	2.9	3.0	2.9	3.2	2.3	2.8	3.2	15-21	0	0

RAND *MG811-T.E.8h*

Table E.9
Officer Predeployment Training, Helpfulness Ratings, by AFSC, Grade, and Deployment Subgroups

Light Blue = Mild Helpfulness Ratings of 3.1–3.4
Medium Blue = Medium Helpfulness Ratings of 3.5–3.9
Dark Blue = High Helpfulness Ratings of 4.0–5.0

2-Digit AFSC	Deployed Location	Grade	Gather/Interpret Info	Respect Cultural Diffs	Apply Social Etiquette	Establish Credibility	Change Behavior	Verbal/Nonverbal Comm	Manage Stress	Negotiate with Others	Apply Regional Knowledge	Resolve Conflict	Self-Initiated Learning	Foreign Lang Skills	Influence Others	Establish Authority	# Respondents	Invited Sample	Population
All Deployed Personnel	**All Deployed Personnel**	**All Deployed Personnel**	**3.0**	**3.3**	**3.1**	**2.8**	**3.0**	**2.4**	**3.1**	**2.6**	**3.1**	**2.6**	**3.2**	**2.0**	**2.5**	**2.6**	**2,390-4,520**	**23,020**	**88,304**
Commander (10XX)	IRAQ/AFG	LO															0	0	0
		HI															2-4	18	18
	OTHER	LO															0	0	0
		HI															3-6	32	32
Pilot (11XX)	IRAQ/AFG	LO	2.9	3.1	3.1	2.2	3.1	1.6	2.9	1.8	3.5	1.9	3.2	1.4	1.8	1.9	10-23	150	700
		HI	2.6	2.8	2.8	2.1	2.6	1.7	2.6	1.9	2.8	2.1	2.9	1.6	2.2	2.2	27-40	150	392
	OTHER	LO	2.5	2.8	2.9	2.4	2.7	1.9	2.6	2.0	2.9	2.3	2.8	1.7	2.1	2.1	18-26	150	2,711
		HI	2.7	3.2	3.0	2.3	2.9	1.9	2.7	2.1	2.8	2.2	3.0	1.4	2.0	2.0	20-39	150	962
Navigator (12XX)	IRAQ/AFG	LO	2.6	2.9	2.6	2.4	2.2	1.6	2.9	2.3	2.9	2.4	3.1	1.4	2.3	2.4	19-29	150	254
		HI	2.8	3.4	2.8	2.3	2.5	2.1	2.8	2.0	3.1	2.6	3.4	1.7	2.4	2.5	9-22	150	161
	OTHER	LO	3.0	3.5	3.0	2.9	3.0	2.5	2.7	2.4	3.3	2.3	3.4	2.2	2.4	2.3	12-23	150	900
		HI	2.6	3.0	2.8	2.3	2.8	1.9	2.5	2.1	2.9	2.0	2.8	1.6	2.0	1.9	33-52	150	372
Space, Missile, and Command and Control (13XX)	IRAQ/AFG	LO	2.6	2.9	2.7	2.6	2.8	2.1	2.9	2.5	3.2	2.4	3.3	1.8	2.3	2.4	14-25	150	153
		HI	1.8	2.9	2.3	2.2	2.6	1.4	2.1	2.1	2.8	2.0	3.1	1.3	2.1	2.1	8-16	70	70
	OTHER	LO	2.9	3.2	2.8	2.8	2.8	2.3	2.7	2.7	3.1	2.7	3.5	2.3	2.5	2.4	12-22	150	413
		HI	2.7	3.3	3.1	2.5	3.1	1.5	2.7	2.1	3.1	1.9	3.2	1.5	2.1	2.0	15-41	150	184
Intelligence (14XX)	IRAQ/AFG	LO	3.3	3.4	3.3	2.6	2.9	2.5	3.0	2.3	3.6	2.2	3.8	2.0	2.2	2.2	13-40	150	236
		HI	3.2	2.8	2.6	2.0	2.8	2.2	2.7	2.0	2.8	1.4	3.9	1.3	2.5	2.5	7-20	68	68
	OTHER	LO	2.8	3.6	3.2	2.3	3.1	1.9	2.9	2.0	3.1	2.1	3.6	1.5	2.1	1.9	13-31	150	427
		HI	2.8	3.4	3.1	2.3	3.1	1.9	2.8	2.2	3.1	2.2	3.4	1.5	2.4	2.1	13-33	108	108
Weather (15XX)	IRAQ/AFG	LO															2-5	37	37
		HI															4-6	15	15
	OTHER	LO															3-11	38	38
		HI															1-3	9	9
International Affairs/FAO (16XX)	IRAQ/AFG	LO															4-8	6	6
		HI	2.3	3.2	2.5	2.8	2.4	1.8	2.7	2.4	2.8	2.4	3.8	1.5	2.5	2.3	11-33	39	39
	OTHER	LO															4-5	5	5
		HI	2.0	3.1	2.3	2.2	2.4	2.0	2.1	1.8	2.4	1.8	3.2	1.7	1.7	1.5	6-13	43	43
Logistics Commander (20XX)	IRAQ/AFG	LO															0	0	0
		HI															3-4	18	18
	OTHER	LO															0	0	0
		HI															3-5	10	10

RAND *MG811-T.E.9a*

Table E.9—continued

2-Digit AFSC	Deployed Location	Grade	Gather/Interpret Info	Respect Cultural Diffs	Apply Social Etiquette	Establish Credibility	Change Behavior	Verbal/Nonverbal Comm	Manage Stress	Negotiate with Others	Apply Regional Knowledge	Resolve Conflict	Self-Initiated Learning	Foreign Lang Skills	Influence Others	Establish Authority	# Respondents	Invited Sample	Population
Aircraft Maintenance (21XX)	IRAQ/AFG	LO	2.8	3.2	3.1	3.1	3.1	2.9	3.4	2.7	2.8	2.8	3.4	2.3	3.1	2.9	10-15	150	295
		HI	2.4	2.9	2.7	2.4	2.3	2.0	2.5	2.3	2.8	2.0	3.1	1.6	2.2	2.2	19-36	133	133
	OTHER	LO	2.7	3.0	3.1	2.7	2.7	2.0	3.1	2.5	2.9	2.2	3.2	1.8	2.1	2.3	12-24	150	470
		HI	2.5	3.4	2.9	2.2	2.6	1.9	2.7	2.1	2.5	1.8	3.1	1.6	1.9	1.8	21-33	150	153
Support Commander (30XX)	IRAQ/AFG	LO															0	0	0
		HI															2-8	26	26
	OTHER	LO															0	0	0
		HI	2.5	3.4	3.0	2.6	2.9	2.2	3.1	2.2	3.3	2.0	3.6	2.0	2.4	2.3	6-8	26	26
Security Forces (31XX)	IRAQ/AFG	LO	3.2	3.2	2.8	2.9	2.8	2.1	3.1	2.7	3.3	3.0	3.5	1.9	3.0	3.1	14-24	150	178
		HI															3-9	45	45
	OTHER	LO															3-8	50	50
		HI															0-1	12	12
Civil Engineer (32XX)	IRAQ/AFG	LO	2.9	3.4	3.0	2.8	3.0	2.3	3.0	2.7	3.1	2.7	3.3	1.6	2.7	2.7	14-28	150	218
		HI	2.5	3.2	3.0	2.6	2.8	1.6	2.8	2.3	2.8	2.3	3.3	1.4	2.4	1.9	18-33	98	98
	OTHER	LO	2.7	3.0	2.9	2.5	2.7	2.3	2.8	2.5	2.9	2.5	3.4	1.8	2.6	2.3	9-19	139	139
		HI	2.9	3.5	3.6	3.3	3.4	2.5	2.9	3.0	3.3	2.6	4.1	1.8	2.8	2.6	5-12	48	48
Comm-Information Systems (33XX)	IRAQ/AFG	LO	2.8	3.1	2.7	2.4	2.7	1.6	2.9	2.1	2.9	2.2	3.6	1.4	2.3	2.1	16-30	150	317
		HI	2.2	3.1	2.7	2.2	2.7	1.7	2.6	2.1	2.8	1.9	3.1	1.5	1.8	1.9	16-27	91	91
	OTHER	LO	3.0	3.4	3.1	2.6	3.1	1.9	3.2	2.1	3.2	2.2	3.5	1.7	2.1	2.2	7-20	150	397
		HI	2.8	3.1	2.7	2.4	3.0	1.7	2.8	2.0	3.2	2.0	3.5	1.4	1.9	1.7	20-33	94	94
Services (34XX)	IRAQ/AFG	LO	2.3	2.8	2.3	2.1	2.3	1.3	2.3	1.7	2.2	2.0	3.4	1.3	2.0	2.0	6-14	45	45
		HI															1-2	15	15
	OTHER	LO	3.1	3.3	3.2	3.1	3.1	2.3	3.1	2.7	3.2	2.7	3.6	2.3	2.7	2.4	8-15	87	87
		HI	3.3	3.0	3.0	2.4	3.0	1.7	2.8	2.2	3.2	2.2	3.0	1.8	2.8	2.0	5-6	18	18
Public Affairs (35XX)	IRAQ/AFG	LO															3-4	32	32
		HI															1-5	11	11
	OTHER	LO	3.3	3.3	3.3	2.7	2.9	2.4	2.8	2.3	3.4	2.3	3.9	1.8	2.1	2.0	6-8	40	40
		HI															0-3	25	26
Personnel (36XX)	IRAQ/AFG	LO															0	26	26
		HI															0	15	15
	OTHER	LO															0	41	41
		HI															0	9	9
Manpower-Personnel (37XX)	IRAQ/AFG	LO	2.1	3.2	2.3	1.9	2.3	1.8	2.5	2.1	2.5	1.9	3.4	1.3	2.1	1.7	6-10	51	51
		HI															4-14	36	36
	OTHER	LO	2.7	3.2	3.0	2.9	2.9	2.0	2.8	2.3	2.7	2.4	3.3	1.8	2.5	2.4	18-35	83	83
		HI															4-10	31	31
Manpower (38XX)	IRAQ/AFG	LO															0	3	3
		HI															0	1	1
	OTHER	LO															0	13	13
		HI															0	3	3

RAND *MG811-T.E.9b*

Table E.9—continued

2-Digit AFSC	Deployed Location	Grade	Gather/Interpret Info	Respect Cultural Diffs	Apply Social Etiquette	Establish Credibility	Change Behavior	Verbal/Nonverbal Comm	Manage Stress	Negotiate with Others	Apply Regional Knowledge	Resolve Conflict	Self-Initiated Learning	Foreign Lang Skills	Influence Others	Establish Authority	# Respondents	Invited Sample	Population
Medical Commander (40XX)	IRAQ/AFG	LO															0	0	0
		HI															2-5	7	7
	OTHER	LO															0	0	0
		HI															2-4	14	14
Health Services Administrator (41XX)	IRAQ/AFG	LO															3-6	40	40
		HI	2.2	2.5	2.5	2.3	2.1	1.8	2.5	1.9	2.5	2.2	3.2	1.6	2.3	2.2	10-15	44	44
	OTHER	LO															3-8	44	44
		HI															3-8	37	37
Biomedical Clinician (42XX)	IRAQ/AFG	LO	2.7	3.1	2.9	2.6	2.4	1.7	2.7	2.5	3.1	2.4	3.3	1.4	2.2	2.0	11-19	70	70
		HI															3-8	34	34
	OTHER	LO															2-3	48	48
		HI															1-3	17	17
Biomedical Specialists (43XX)	IRAQ/AFG	LO															3-4	32	32
		HI	2.7	3.2	2.8	2.6	2.8	2.1	2.7	2.3	2.7	2.3	3.0	1.7	2.5	2.3	7-17	49	49
	OTHER	LO															2-11	34	34
		HI	3.1	3.3	3.2	3.0	3.1	2.5	2.8	2.8	3.1	2.5	3.5	1.8	2.6	2.7	5-15	42	42
Physician (44XX)	IRAQ/AFG	LO															1-3	46	46
		HI	2.5	3.1	2.8	2.5	2.6	1.7	2.9	1.9	3.3	2.1	3.4	1.3	2.2	2.2	10-20	120	120
	OTHER	LO															0-3	48	48
		HI	2.9	3.1	2.6	2.4	2.8	1.8	2.7	2.1	2.6	2.0	3.2	1.6	1.9	1.8	10-19	77	77
Surgery (45XX)	IRAQ/AFG	LO															0-1	16	16
		HI	2.5	3.4	2.9	3.2	3.0	2.4	2.7	2.4	2.8	2.8	3.5	2.2	2.5	2.8	6-14	113	113
	OTHER	LO															0	4	4
		HI	3.2	3.8	3.5	3.3	3.5	2.2	3.2	1.8	3.0	2.0	3.6	1.4	1.6	1.8	5-7	38	38
Nurse (46XX)	IRAQ/AFG	LO	2.9	3.5	3.2	2.8	3.0	1.9	3.0	2.3	3.0	2.6	3.2	1.8	2.3	2.1	16-34	150	258
		HI	2.3	3.1	2.6	2.2	2.5	1.7	2.6	1.9	2.6	2.0	3.1	1.5	2.0	1.8	23-37	99	99
	OTHER	LO	3.0	3.1	2.8	2.8	3.0	2.3	2.8	2.4	3.0	2.3	3.4	2.0	2.4	2.5	22-31	150	269
		HI	2.9	3.4	3.0	2.6	2.8	2.0	2.8	2.2	3.0	2.3	3.7	1.6	2.1	2.0	19-30	94	94
Dental (47XX)	IRAQ/AFG	LO															0	4	4
		HI															0-3	12	12
	OTHER	LO															0-1	8	8
		HI															4-8	24	24
Aerospace Medicine (48XX)	IRAQ/AFG	LO															0-2	26	26
		HI	3.4	3.7	3.6	3.5	3.6	2.1	3.6	2.8	3.4	2.7	3.8	1.3	2.9	2.9	6-14	51	51
	OTHER	LO															2-4	57	57
		HI	2.8	2.9	3.0	2.2	2.9	2.1	2.5	2.1	3.2	1.8	3.2	1.2	1.8	1.8	13-22	81	81

RAND *MG811-T.E.9c*

Table E.9—continued

2-Digit AFSC	Deployed Location	Grade	Gather/Interpret Info	Respect Cultural Diffs	Apply Social Etiquette	Establish Credibility	Change Behavior	Verbal/Nonverbal Comm	Manage Stress	Negotiate with Others	Apply Regional Knowledge	Resolve Conflict	Self-Initiated Learning	Foreign Lang Skills	Influence Others	Establish Authority	# Respondents	Invited Sample	Population
Law (51XX)	IRAQ/AFG	LO															3-7	56	56
		HI	2.4	3.4	3.3	2.3	2.3	1.6	3.0	2.5	3.0	2.5	3.5	1.3	2.3	2.2	7-16	43	43
	OTHER	LO															1-3	37	37
		HI	2.3	3.0	2.6	2.6	2.5	1.7	2.4	2.5	2.4	2.6	3.1	1.6	2.4	2.3	9-14	49	49
Chaplain (52XX)	IRAQ/AFG	LO	3.3	3.7	3.5	3.0	3.5	2.0	3.2	2.3	3.8	2.4	3.7	1.2	2.6	2.3	6-14	43	43
		HI	2.9	3.5	3.5	3.0	3.2	1.2	3.1	2.2	3.2	2.6	3.8	1.2	2.3	2.3	5-18	49	49
	OTHER	LO	3.1	3.5	3.1	3.1	3.2	2.2	3.2	2.8	3.2	2.2	3.6	1.8	2.6	2.2	5-11	43	43
		HI															3-9	25	25
Program Director (60XX)	IRAQ/AFG	LO															0	0	0
		HI															0	1	1
	OTHER	LO															0	0	0
		HI															0	0	0
Scientific/Research (61XX)	IRAQ/AFG	LO															0-2	9	9
		HI															0-3	4	4
	OTHER	LO															4-5	15	15
		HI															2-3	16	16
Developmental Engineering (62XX)	IRAQ/AFG	LO															1-3	21	21
		HI															0-1	10	10
	OTHER	LO	2.0	2.8	2.3	1.6	2.2	2.2	1.8	1.6	2.4	1.6	3.3	1.4	1.6	1.6	5-6	31	31
		HI															0	5	5
Acquisition (63XX)	IRAQ/AFG	LO															3-6	33	33
		HI															4-7	42	42
	OTHER	LO															2	31	31
		HI															3	17	17
Contracting (64XX)	IRAQ/AFG	LO	2.7	3.4	3.1	2.8	2.6	1.7	2.8	3.0	3.2	2.6	3.4	1.3	2.5	2.3	12-23	111	111
		HI	2.7	3.3	3.4	3.1	3.1	2.4	2.7	3.1	3.5	2.4	3.5	1.4	2.7	2.8	7-17	52	52
	OTHER	LO	3.2	3.6	3.6	3.6	3.1	2.8	3.3	4.1	3.3	3.1	3.8	2.0	3.3	3.1	5-17	69	69
		HI															1-2	4	4
Finance (65XX)	IRAQ/AFG	LO															2-7	43	43
		HI															2-4	37	37
	OTHER	LO															1-3	22	22
		HI															4-7	17	17
Special Investigations (71XX)	IRAQ/AFG	LO															4	30	30
		HI															2-4	14	14
	OTHER	LO															3-4	17	17
		HI															0-1	4	4

RAND *MG811-T.E.9d*

Table E.9—continued

2-Digit AFSC	Deployed Location	Grade	Gather/Interpret Info	Respect Cultural Diffs	Apply Social Etiquette	Establish Credibility	Change Behavior	Verbal/Nonverbal Comm	Manage Stress	Negotiate with Others	Apply Regional Knowledge	Resolve Conflict	Self-Initiated Learning	Foreign Lang Skills	Influence Others	Establish Authority	# Respondents	Invited Sample	Population
Commander—USAFA Cadet Squadron (80XX)	IRAQ/AFG	LO															0	0	0
		HI															0	0	0
	OTHER	LO															0	0	0
		HI															0	1	1
Instructor (81XX)	IRAQ/AFG	LO															1	4	4
		HI															0-3	8	8
	OTHER	LO															0	6	6
		HI															0	8	8
Academic Program Manager (82XX)	IRAQ/AFG	LO															0	1	1
		HI															0	3	3
	OTHER	LO															0	1	1
		HI															0	3	3
Recruiting Service (83XX)	IRAQ/AFG	LO															0	4	4
		HI															0	0	0
	OTHER	LO															0	2	2
		HI															0	0	0
Historian (84XX)	IRAQ/AFG	LO															0	0	0
		HI															0	0	0
	OTHER	LO															0	0	0
		HI															0	0	0
USAF Honor Guard (85XX)	IRAQ/AFG	LO															0	0	0
		HI															0	0	0
	OTHER	LO															0	0	0
		HI															0	1	1
Command and Control (86XX)	IRAQ/AFG	LO															0	2	2
		HI															1-2	3	3
	OTHER	LO															1-2	3	3
		HI															0	5	5
Inspector General (87XX)	IRAQ/AFG	LO															0	0	0
		HI															0	0	0
	OTHER	LO															0	0	0
		HI															0	1	1
Protocol/Aide (88XX)	IRAQ/AFG	LO															0-2	1	1
		HI															0	0	0
	OTHER	LO															1-3	0	0
		HI															0	0	0

RAND *MG811-T.E.9e*

Table E.9—continued

2-Digit AFSC	Deployed Location	Grade	Gather/Interpret Info	Respect Cultural Diffs	Apply Social Etiquette	Establish Credibility	Change Behavior	Verbal/Nonverbal Comm	Manage Stress	Negotiate with Others	Apply Regional Knowledge	Resolve Conflict	Self-Initiated Learning	Foreign Lang Skills	Influence Others	Establish Authority	# Respondents	Invited Sample	Population
General Officer (90XX)	IRAQ/AFG	LO															0	0	0
		HI															1-2	10	10
	OTHER	LO															0	0	0
		HI															0	11	11
Commander (91XX)	IRAQ/AFG	LO															0	0	0
		HI															1-2	13	13
	OTHER	LO															0	0	0
		HI															4-6	18	18
Student (92XX)	IRAQ/AFG	LO															0	22	22
		HI															0	7	7
	OTHER	LO															0-1	65	65
		HI															0	12	12
Patient (93XX)	IRAQ/AFG	LO															0	0	0
		HI															0	0	0
	OTHER	LO															0	0	0
		HI															0	0	0
Nuclear Weapons Custodian (94XX)	IRAQ/AFG	LO															0	0	0
		HI															0	0	0
	OTHER	LO															0	0	0
		HI															0	0	0
Non-extended Active Duty (95XX)	IRAQ/AFG	LO															0	0	0
		HI															0	0	0
	OTHER	LO															0	0	0
		HI															0	1	1
Unclassified Officer (96XX)	IRAQ/AFG	LO															0	0	0
		HI															0	0	0
	OTHER	LO															0-1	2	2
		HI															0	0	0
Executive Officer above Wing Level (97XX)	IRAQ/AFG	LO															1	3	3
		HI															0-1	12	12
	OTHER	LO															0	6	6
		HI															0	7	7

RAND *MG811-T.E.9f*

Table E.10
Enlisted Predeployment Training, Helpfulness Ratings, by AFSC, Grade, and Deployment Subgroups

Light Blue = Mild Helpfulness Ratings of 3.1–3.4 | Medium Blue = Medium Helpfulness Ratings of 3.5–3.9 | Dark Blue = High Helpfulness Ratings of 4.0–5.0

2-Digit AFSC	Deployed Location	Grade	Gather/Interpret	Respect Cultural Diffs	Apply Social Etiquette	Establish Credibility	Change Behavior	Verbal/Nonverbal Comm	Manage Stress	Negotiate with Others	Apply Regional Knowledge	Resolve Conflict	Self-Initiated Learning	Foreign Lang Skills	Influence Others	Establish Authority	# Respondents	Invited Sample	Population
All Deployed Personnel	All Deployed Personnel	All Deployed Personnel	3.0	3.3	3.1	2.8	3.0	2.4	3.1	2.6	3.1	2.6	3.2	2.0	2.5	2.6	2,390-4,520	23,020	88,304
Aircrew Operations (1AXXX)	IRAQ/AFG	LO	2.8	2.8	3.2	2.8	2.9	2.4	3.1	2.7	3.1	2.7	3.0	2.1	2.6	2.7	9-15	150	218
		HI	3.1	3.5	3.5	3.2	3.4	2.7	3.3	3.1	3.5	2.8	3.7	2.0	3.1	2.8	10-24	150	633
	OTHER	LO	3.0	3.5	3.4	3.0	3.0	2.6	3.1	3.1	3.3	2.6	3.1	2.0	2.7	2.6	12-21	150	1,248
		HI	2.7	3.0	3.4	2.6	3.1	2.2	3.1	2.3	3.1	2.4	2.8	2.0	2.1	2.2	17-36	150	1,431
Command & Control Systems Operations (1CXXX)	IRAQ/AFG	LO	2.6	2.8	2.5	2.3	2.5	1.8	2.5	2.3	2.9	1.9	2.6	1.8	2.2	2.1	10-16	150	859
		HI	2.7	3.1	2.7	2.6	2.7	2.3	2.9	2.4	2.9	2.4	3.0	1.9	2.3	2.4	26-39	150	983
	OTHER	LO	3.1	3.3	3.5	2.9	3.1	2.8	3.2	2.8	3.3	2.5	3.1	2.5	2.7	2.6	11-18	150	554
		HI	3.0	3.2	3.2	2.8	2.9	2.5	3.2	2.6	3.2	2.7	3.4	2.1	2.3	2.6	26-47	150	676
Intelligence (1NXXX)	IRAQ/AFG	LO	2.5	3.5	2.8	2.4	2.4	1.9	2.3	2.2	2.9	2.5	3.8	1.8	2.5	2.7	8-14	150	290
		HI	3.0	3.2	2.8	2.6	2.6	2.4	2.9	2.6	3.5	2.4	3.7	1.8	2.7	2.6	18-46	150	313
	OTHER	LO	3.4	3.6	3.4	2.3	3.0	2.1	2.8	2.6	3.8	2.0	3.8	1.4	1.9	2.0	5-17	150	436
		HI	2.8	3.4	3.2	2.6	3.0	1.9	2.9	2.4	3.2	2.1	3.6	1.6	2.3	2.3	20-41	150	437
Safety (1SXXX)	IRAQ/AFG	LO															0	0	0
		HI															2-11	39	39
	OTHER	LO															0	1	1
		HI															2-9	41	41
Aircrew Protection (1TXXX)	IRAQ/AFG	LO	3.0	3.2	3.5	3.2	3.2	2.8	3.2	2.4	3.2	2.6	3.2	2.4	2.4	2.6	5-6	115	115
		HI	3.2	3.6	3.4	2.9	3.4	2.9	3.6	2.9	3.7	2.9	3.9	2.2	3.0	3.0	9-17	118	118
	OTHER	LO	3.1	3.1	3.1	3.1	2.9	1.8	3.0	2.6	3.0	2.4	3.5	1.6	2.3	2.0	5-9	150	276
		HI	3.0	3.3	3.1	3.2	3.1	2.5	3.2	2.7	3.2	2.7	3.6	2.2	2.5	2.5	17-35	150	290
Weather (1WXXX)	IRAQ/AFG	LO															3-5	68	68
		HI	2.7	3.2	2.9	2.2	2.8	2.0	2.8	2.2	3.2	2.2	3.3	1.7	2.4	2.3	20-44	150	217
	OTHER	LO															2-3	54	54
		HI	2.8	3.3	3.2	2.6	3.1	2.3	3.1	2.2	3.6	2.0	3.3	1.9	1.8	2.1	16-35	127	127
Manned Aerospace Maintenance (2AXXX)	IRAQ/AFG	LO															1-7	150	2,505
		HI	2.7	3.3	3.1	2.6	3.0	2.1	3.2	2.3	3.0	2.3	3.0	1.7	2.3	2.5	19-45	150	2,396
	OTHER	LO															4-10	150	5,727
		HI	2.8	3.1	3.1	2.5	2.9	2.4	3.1	2.3	3.0	2.3	3.1	2.3	2.3	2.2	27-40	150	5,923

RAND *MG811-T.E.10a*

Table E.10—continued

2-Digit AFSC	Deployed Location	Grade	Gather/Interpret	Respect Cultural Diffs	Apply Social Etiquette	Establish Credibility	Change Behavior	Verbal/Nonverbal Comm	Manage Stress	Negotiate with Others	Apply Regional Knowledge	Resolve Conflict	Self-Initiated Learning	Foreign Lang Skills	Influence Others	Establish Authority	# Respondents	Invited Sample	Population
Communications & Electronics (2EXXX)	IRAQ/AFG	LO	3.0	3.1	3.1	2.6	3.2	2.5	3.3	2.6	3.2	2.7	3.2	2.4	2.7	2.8	**11-21**	150	705
		HI	2.2	2.9	2.6	2.3	2.5	1.9	2.7	2.1	2.8	2.2	3.0	1.7	2.2	2.2	**32-56**	150	596
	OTHER	LO	3.3	3.4	3.2	2.9	3.1	2.5	3.5	2.6	3.4	3.1	3.3	2.3	2.5	2.8	**7-25**	150	654
		HI	2.9	3.4	3.1	2.8	3.3	2.3	3.0	2.2	3.1	2.1	3.3	1.8	2.2	2.2	**24-42**	150	689
Fuels (2FXXX)	IRAQ/AFG	LO	2.9	3.3	2.9	2.9	2.9	2.2	3.3	2.8	3.1	3.1	3.3	2.3	2.8	3.1	**11-21**	150	209
		HI	2.8	3.2	3.0	2.7	2.8	2.2	3.2	2.6	3.2	2.6	3.2	1.9	2.6	2.7	**22-45**	150	264
	OTHER	LO	3.0	3.4	3.2	2.8	3.2	2.9	2.8	2.7	3.6	2.8	3.3	2.5	2.4	2.5	**8-13**	150	668
		HI	2.9	3.3	3.1	2.9	3.0	2.4	3.0	2.7	3.2	2.7	3.3	2.1	2.6	2.7	**27-53**	150	709
Logistics Plans (2GXXX)	IRAQ/AFG	LO															**0-2**	22	22
		HI	2.6	3.0	2.5	2.1	2.6	1.9	2.8	2.1	2.5	2.1	2.4	1.8	2.1	2.1	**16-21**	87	87
	OTHER	LO															**3-6**	37	37
		HI	2.9	3.4	3.0	2.4	3.3	2.1	3.3	2.2	3.1	2.3	3.2	1.6	2.2	2.1	**14-30**	119	119
Missile & Space Systems Maintenance (2MXXX)	IRAQ/AFG	LO															**0-2**	59	59
		HI															**2-3**	22	22
	OTHER	LO															**0**	39	39
		HI															**1**	7	7
Precision Measurement Equipment Lab (2PXXX)	IRAQ/AFG	LO															**0**	5	5
		HI															**0-2**	5	5
	OTHER	LO															**1-3**	42	42
		HI	2.6	3.4	2.7	2.4	2.8	2.2	2.9	2.1	2.7	2.1	3.1	1.8	2.0	2.2	**11-21**	61	61
Maintenance Management Systems (2RXXX)	IRAQ/AFG	LO															**1-4**	42	42
		HI															**2-10**	06	06
	OTHER	LO															**2-6**	92	92
		HI	3.0	3.2	3.4	2.6	3.2	2.5	3.2	2.4	3.1	2.4	3.1	2.0	2.3	2.3	**13-24**	100	100
Supply (2SXXX)	IRAQ/AFG	LO	2.9	3.4	2.6	2.0	3.2	2.2	3.7	2.3	2.4	2.0	2.9	1.9	2.4	2.4	**7-11**	150	374
		HI	3.0	3.3	3.3	2.9	2.9	2.6	3.1	2.6	3.2	2.6	3.2	2.1	2.5	2.5	**17-32**	150	643
	OTHER	LO	3.3	3.5	3.4	2.5	3.5	2.9	3.5	2.6	3.4	2.6	2.8	2.0	2.6	2.9	**6-11**	150	650
		HI	2.8	3.2	3.1	2.7	2.8	2.2	2.9	2.3	2.9	2.4	3.0	1.8	2.3	2.4	**19-30**	150	1,067
Transportation & Vehicle Maintenance (2TXXX)	IRAQ/AFG	LO	3.2	3.3	3.2	3.0	3.1	2.8	3.3	3.0	3.3	2.9	3.1	2.7	3.1	3.2	**13-20**	150	1,232
		HI	2.8	3.0	3.0	2.8	2.8	2.1	2.9	2.4	2.8	2.6	2.9	1.9	2.5	2.6	**31-54**	150	1,219
	OTHER	LO	3.1	3.1	3.1	2.8	2.8	3.2	3.4	3.1	3.4	3.0	3.2	2.3	2.6	2.8	**7-11**	150	1,207
		HI	3.1	3.4	3.1	2.8	3.3	2.4	3.1	2.5	3.3	2.6	3.2	2.2	2.6	2.7	**22-41**	150	1,275
Munitions & Weapons (2WXXX)	IRAQ/AFG	LO	2.7	3.3	3.2	2.5	2.8	1.9	3.0	2.6	2.9	2.7	2.9	1.9	2.5	2.4	**12-19**	150	768
		HI	2.8	2.9	2.8	2.6	2.9	2.2	3.0	2.3	3.1	2.6	3.0	2.1	2.4	2.7	**17-33**	150	714
	OTHER	LO	3.5	3.7	3.5	3.6	3.1	2.8	3.5	3.4	3.7	2.8	3.5	2.7	3.0	3.1	**9-16**	150	980
		HI	2.8	3.0	3.0	2.8	2.9	2.2	2.9	2.4	2.9	2.5	3.1	2.2	2.4	2.4	**26-42**	150	905

RAND *MG811-T.E.10b*

Table E.10—continued

2-Digit AFSC	Deployed Location	Grade	Gather/Interpret	Respect Cultural Diffs	Apply Social Etiquette	Establish Credibility	Change Behavior	Verbal/Nonverbal Comm	Manage Stress	Negotiate with Others	Apply Regional Knowledge	Resolve Conflict	Self-Initiated Learning	Foreign Lang Skills	Influence Others	Establish Authority	# Respondents	Invited Sample	Population
Information Management (3AXXX)	IRAQ/AFG	LO	2.9	3.3	3.4	2.7	3.1	2.5	3.3	2.6	3.1	2.9	3.1	2.2	3.0	2.9	10-16	150	220
		HI	2.8	3.5	3.0	2.7	3.2	2.2	3.2	2.3	3.1	3.0	3.3	1.9	2.6	2.9	19-43	150	539
	OTHER	LO	3.2	3.4	3.4	2.9	3.1	2.7	3.5	3.0	3.2	2.9	3.3	2.3	2.7	2.7	8-18	150	329
		HI	3.3	3.7	3.4	2.9	3.3	2.4	3.4	2.4	3.5	2.2	3.7	1.9	2.3	2.2	14-36	150	756
Communications & Computer Systems (3CXXX)	IRAQ/AFG	LO	2.8	3.0	2.8	2.7	2.7	2.0	2.8	2.7	2.8	2.8	2.9	1.9	2.6	2.8	12-18	150	405
		HI	2.9	3.4	3.0	2.8	2.6	2.1	3.2	2.4	3.1	2.6	3.3	1.9	2.7	2.4	12-28	150	627
	OTHER	LO	3.3	3.5	2.9	2.9	3.3	2.1	2.9	2.6	3.3	2.5	3.0	2.1	2.4	2.2	7-11	150	617
		HI	2.8	3.2	2.8	2.5	2.9	2.1	3.2	2.3	2.8	2.3	3.1	1.6	2.3	2.2	19-35	150	1,026
Civil Engineering (3EXXX)	IRAQ/AFG	LO	3.0	3.1	3.3	2.7	2.9	2.3	3.1	2.4	2.9	2.7	3.3	2.0	2.4	2.6	8-15	150	1,839
		HI	3.1	3.3	3.2	2.9	3.0	2.4	3.0	2.9	3.3	2.9	3.4	2.1	2.9	2.9	31-53	150	1,826
	OTHER	LO															4-12	150	2,039
		HI	3.1	3.4	3.1	3.0	3.2	2.1	3.1	2.4	3.2	2.6	3.5	2.0	2.7	2.6	18-34	150	1,919
Historian (3HXXX)	IRAQ/AFG	LO															0	0	0
		HI															0-1	5	5
	OTHER	LO															0	0	0
		HI															2	7	7
Services (3MXXX)	IRAQ/AFG	LO	3.4	3.4	3.4	2.9	3.0	2.4	3.5	2.9	3.4	3.0	3.3	1.9	2.9	3.1	7-22	150	208
		HI	2.8	3.2	3.1	2.6	2.9	2.4	2.9	2.6	2.8	2.7	2.9	2.2	2.6	2.7	18-37	150	268
	OTHER	LO	3.2	3.7	3.1	2.8	3.3	2.6	3.6	2.3	2.9	2.5	3.1	2.2	2.7	2.5	6-14	150	703
		HI	3.1	3.5	3.4	3.2	3.1	2.5	3.3	2.9	3.2	2.9	3.4	2.4	2.6	2.7	25-45	150	762
Public Affairs (3NXXX)	IRAQ/AFG	LO															1-4	33	33
		HI	3.1	3.2	3.3	2.5	3.1	2.0	3.2	2.4	3.1	2.5	3.5	1.5	2.5	2.5	10-25	78	78
	OTHER	LO															1-2	20	20
		HI	2.8	3.4	3.1	2.7	2.7	2.4	3.0	2.1	2.8	2.7	3.9	1.8	2.3	2.5	6-15	77	77
Security Forces (Military Police) (3PXXX)	IRAQ/AFG	LO	3.6	3.9	3.2	3.6	2.9	2.2	3.4	3.2	3.0	3.5	3.4	1.7	2.6	3.5	9-12	150	3,442
		HI	3.4	3.5	3.1	3.4	3.0	2.6	3.3	3.2	3.5	3.4	3.4	2.0	3.1	3.5	15-33	150	2,206
	OTHER	LO	3.3	3.7	3.5	3.2	3.3	2.6	3.5	3.2	3.3	3.1	3.6	2.6	3.2	3.3	9-14	150	2,508
		HI	3.1	3.3	3.0	3.1	2.8	2.6	3.0	2.8	3.3	2.9	3.3	2.3	2.8	2.9	20-35	150	1,633
Mission Support (3SXXX)	IRAQ/AFG	LO															3-7	92	92
		HI	2.8	3.1	2.8	2.8	3.0	2.1	3.1	2.5	3.0	2.6	3.4	1.9	2.5	2.6	15-38	150	345
	OTHER	LO	4.0	3.7	3.8	3.4	3.8	3.3	3.8	3.5	3.8	2.9	4.0	2.6	2.7	2.7	8-14	121	121
		HI	3.0	3.2	2.8	2.7	2.7	2.2	2.9	2.1	2.9	2.4	3.3	1.8	2.1	2.1	27-42	150	467
Visual Information (3VXXX)	IRAQ/AFG	LO															1-3	54	54
		HI	2.8	3.3	3.1	2.9	2.9	2.4	2.8	2.6	2.9	2.6	3.6	1.8	2.8	2.7	12-20	81	81
	OTHER	LO															1-2	35	35
		HI	3.4	3.5	3.2	3.1	3.3	2.5	3.5	2.9	3.4	2.7	3.4	2.5	2.7	2.7	6-15	60	60

RAND *MG811-T.E.10c*

Table E.10—continued

2-Digit AFSC	Deployed Location	Grade	Gather/Interpret	Respect Cultural Diffs	Apply Social Etiquette	Establish Credibility	Change Behavior	Verbal/Nonverbal Comm	Manage Stress	Negotiate with Others	Apply Regional Knowledge	Resolve Conflict	Self-Initiated Learning	Foreign Lang Skills	Influence Others	Establish Authority	# Respondents	Invited Sample	Population
Medical (4AXXX)	IRAQ/AFG	LO	3.4	3.6	3.4	2.8	3.1	2.6	3.0	2.6	3.4	2.8	3.3	2.1	2.6	2.8	**9-15**	150	157
		HI	3.1	3.5	3.2	3.0	3.2	2.5	3.4	2.6	3.3	2.7	3.4	2.1	2.8	2.7	**11-41**	150	189
	OTHER	LO	3.1	3.4	3.2	3.0	3.0	2.1	3.4	2.8	2.9	3.0	3.8	2.0	2.8	2.7	**7-18**	150	155
		HI	3.3	3.3	3.5	2.6	3.4	2.3	3.3	2.5	3.3	2.5	3.3	2.0	2.3	2.4	**15-28**	150	250
Bioenvironmental Engineering (4BXXX)	IRAQ/AFG	LO															**2**	8	12
		HI															**3-6**	14	25
	OTHER	LO															**0**	16	25
		HI															**2-3**	19	39
Mental Health Services (4CXXX)	IRAQ/AFG	LO															**1-3**	15	21
		HI															**2-7**	22	40
	OTHER	LO															**2**	22	30
		HI															**2-4**	12	23
Diet Therapy (4DXXX)	IRAQ/AFG	LO															**0**	13	22
		HI															**1**	8	13
	OTHER	LO															**0**	5	9
		HI															**0**	1	1
Public Health (4EXXX)	IRAQ/AFG	LO															**0-1**	8	14
		HI															**2**	10	21
	OTHER	LO															**2-3**	17	23
		HI															**2-4**	17	47
Cardiopulmonary Laboratory (4HXXX)	IRAQ/AFG	LO															**2-3**	29	39
		HI															**2-4**	8	27
	OTHER	LO															**1-2**	13	18
		HI															**3-5**	10	27
Physical Medicine (4JXXX)	IRAQ/AFG	LO															**0**	4	4
		HI															**2**	10	24
	OTHER	LO															**0**	0	0
		HI															**0**	3	4
Aerospace Physiology (4MXXX)	IRAQ/AFG	LO															**0**	2	5
		HI															**2**	2	11
	OTHER	LO															**1-3**	16	20
		HI															**4-6**	13	25
Medical Service (4NXXX)	IRAQ/AFG	LO	2.7	3.5	3.1	2.5	2.8	1.7	3.3	2.3	3.2	2.3	2.6	1.6	2.1	2.5	**9-13**	150	271
		HI	2.8	3.2	3.3	2.9	3.1	2.0	3.0	2.4	3.1	2.5	3.3	2.0	2.6	2.6	**22-40**	150	468
	OTHER	LO	2.8	3.4	3.1	2.3	2.8	2.3	3.2	2.3	2.9	2.1	3.4	1.8	2.3	2.2	**9-15**	150	274
		HI	2.6	2.9	2.6	2.5	2.8	2.0	2.8	2.1	2.9	2.4	3.6	1.9	2.3	2.3	**20-37**	150	532

RAND *MG811-T.E.10d*

Table E.10—continued

2-Digit AFSC	Deployed Location	Grade	Gather/Interpret	Respect Cultural Diffs	Apply Social Etiquette	Establish Credibility	Change Behavior	Verbal/Nonverbal Comm	Manage Stress	Negotiate with Others	Apply Regional Knowledge	Resolve Conflict	Self-Initiated Learning	Foreign Lang Skills	Influence Others	Establish Authority	# Respondents	Invited Sample	Population
Pharmacy (4PXXX)	IRAQ/AFG	LO															1-2	14	20
		HI															2	14	32
	OTHER	LO															1	6	9
		HI															4-6	23	36
Diagnostic Imaging (4RXXX)	IRAQ/AFG	LO															2-3	21	30
		HI	1.8	2.7	2.3	2.7	2.3	1.7	2.5	2.0	2.3	2.7	2.5	1.7	2.3	2.5	6	17	28
	OTHER	LO															0	9	14
		HI															1-2	11	30
Medical Laboratory (4TXXX)	IRAQ/AFG	LO															1-2	14	26
		HI															3-6	23	53
	OTHER	LO															1-3	17	27
		HI															2-4	14	32
Optometry (4VXXX)	IRAQ/AFG	LO															0	2	2
		HI															0	3	5
	OTHER	LO															1	3	8
		HI															0	3	8
Dental (4YXXX)	IRAQ/AFG	LO															1-2	20	30
		HI															1-3	19	36
	OTHER	LO															2	26	38
		HI															4-5	24	37
Paralegal (5JXXX)	IRAQ/AFG	LO															0	1	1
		HI	2.6	3.2	3.0	2.8	2.9	1.8	3.1	2.4	2.8	2.5	3.1	1.9	2.4	2.2	8-20	67	67
	OTHER	LO															0-1	6	6
		HI															4-10	44	44
Chaplain Assistant (5RXXX)	IRAQ/AFG	LO															1-3	24	24
		HI															4-11	48	48
	OTHER	LO															3-4	21	21
		HI	3.4	3.7	3.5	3.3	3.5	3.1	3.4	3.3	3.6	2.9	3.6	2.4	2.9	3.3	7-14	33	33
Contracting (6CXXX)	IRAQ/AFG	LO	3.3	3.6	3.5	3.4	3.5	2.0	3.1	3.4	3.2	2.5	3.2	1.8	3.0	3.1	5-10	51	51
		HI	3.2	3.5	3.2	3.4	3.1	2.1	3.0	3.6	3.1	2.8	3.7	1.7	2.6	2.9	22-50	150	211
	OTHER	LO															4-7	59	59
		HI	3.6	3.7	3.3	3.8	3.4	2.7	3.4	4.0	3.4	3.4	3.8	1.8	2.9	3.0	16-39	150	168
Financial (6FXXX)	IRAQ/AFG	LO															3-4	55	55
		HI	2.7	3.5	2.5	2.9	2.7	1.8	2.9	2.8	2.8	2.7	3.8	1.8	2.6	2.9	12-19	118	118
	OTHER	LO															3-4	103	103
		HI	2.7	2.9	2.6	2.7	2.4	2.0	2.6	2.4	2.7	2.2	3.6	1.5	2.4	2.2	20-36	150	156

RAND *MG811-T.E.10e*

Table E.10—continued

2-Digit AFSC	Deployed Location	Grade	Gather/Interpret	Respect Cultural Diffs	Apply Social Etiquette	Establish Credibility	Change Behavior	Verbal/Nonverbal Comm	Manage Stress	Negotiate with Others	Apply Regional Knowledge	Resolve Conflict	Self-Initiated Learning	Foreign Lang Skills	Influence Others	Establish Authority	# Respondents	Invited Sample	Population
Special Investigations (OSI) (7SXXX)	IRAQ/AFG	LO															0	0	0
		HI	4.0	4.1	3.7	4.0	3.6	2.7	3.7	3.2	3.6	3.2	3.6	1.9	3.5	3.0	12-21	121	121
	OTHER	LO															0	0	0
		HI	3.5	3.3	3.4	3.6	3.6	2.0	3.6	3.4	3.6	3.2	3.7	1.8	3.3	3.3	5-7	81	81
Enlisted Aide (8AXXX)	IRAQ/AFG	LO															0	0	0
		HI															0	1	1
	OTHER	LO															0	0	0
		HI															0	9	9
Military Training (8BXXX)	IRAQ/AFG	LO															0	1	1
		HI															2-3	16	16
	OTHER	LO															0	0	0
		HI															1-3	15	15
Family Support Center (8CXXX)	IRAQ/AFG	LO															0	0	0
		HI															0	0	0
	OTHER	LO															0	0	0
		HI															2	6	6
Linguist Debriefer (8DXXX)	IRAQ/AFG	LO															0	0	0
		HI															4-6	11	11
	OTHER	LO															0	0	0
		HI															0	0	0
First Sergeant (8FXXX)	IRAQ/AFG	LO															0	0	0
		HI	2.5	3.2	2.8	2.5	2.8	1.9	2.9	2.2	2.8	2.4	3.1	1.8	2.4	2.4	23-41	141	141
	OTHER	LO															0	0	0
		HI	2.6	3.3	3.0	2.6	2.8	2.1	2.6	2.1	2.8	2.3	3.4	1.8	2.0	2.1	20-34	150	189
USAF Honor Guard (8GXXX)	IRAQ/AFG	LO															0	0	0
		HI															0	1	1
	OTHER	LO															0	0	0
		HI															0	0	0
Postal Specialist (8MXXX)	IRAQ/AFG	LO															0	22	22
		HI	2.8	3.2	3.0	2.7	3.3	2.2	3.1	2.3	3.1	2.6	2.6	2.2	2.4	2.6	5-10	52	52
	OTHER	LO	3.6	3.3	3.6	3.2	3.0	2.9	3.5	3.4	3.5	3.2	3.5	2.4	2.9	3.1	5-11	89	89
		HI	2.8	3.2	2.7	3.0	2.9	2.3	2.9	2.7	2.9	2.5	3.5	2.1	2.7	2.8	17-28	100	100
Recruiter (8RXXX)	IRAQ/AFG	LO															0	3	3
		HI															0-1	6	6
	OTHER	LO															0	4	4
		HI															0-3	15	15

RAND *MG811-T.E.10f*

Table E.10—continued

2-Digit AFSC	Deployed Location	Grade	Gather/Interpret	Respect Cultural Diffs	Apply Social Etiquette	Establish Credibility	Change Behavior	Verbal/Nonverbal Comm	Manage Stress	Negotiate with Others	Apply Regional Knowledge	Resolve Conflict	Self-Initiated Learning	Foreign Lang Skills	Influence Others	Establish Authority	# Respondents	Invited Sample	Population
Professional Military Education Instructor (8TXXX)	IRAQ/AFG	LO															1	0	0
		HI															0	1	1
	OTHER	LO															0	0	0
		HI															0	3	3
Awaiting Retraining (9AXXX)	IRAQ/AFG	LO															1	1	1
		HI															0	0	0
	OTHER	LO															0	4	4
		HI															0	0	0
Dorm Manager (9DXXX)	IRAQ/AFG	LO															0	0	0
		HI															0	9	9
	OTHER	LO															0	1	1
		HI															0	10	10
Senior Enlisted Advisor (9EXXX)	IRAQ/AFG	LO															0	0	0
		HI															0	1	1
	OTHER	LO															0	0	0
		HI															1-3	4	4
Group Superintendent (9GXXX)	IRAQ/AFG	LO															0	0	0
		HI	2.6	3.3	3.4	2.3	2.7	2.0	2.6	2.0	3.3	2.2	3.5	1.8	2.0	2.0	6-8	25	25
	OTHER	LO															0	0	0
		HI															4-8	23	23
Interpreter/Translator (9LXXX)	IRAQ/AFG	LO															0	0	0
		HI															0	0	0
	OTHER	LO															0	0	0
		HI															0	3	3
Technical Applications Specialist (9SXXX)	IRAQ/AFG	LO															0	0	0
		HI															0-1	0	0
	OTHER	LO															0	5	5
		HI															1-2	18	18
Basic Enlisted Trainee (9TXXX)	IRAQ/AFG	LO															0	13	13
		HI															0	0	0
	OTHER	LO															0-1	8	8
		HI															0-1	0	0
Unallotted Airman (9UXXX)	IRAQ/AFG	LO															4-11	0	0
		HI	3.5	3.1	2.9	3.2	3.0	2.3	2.8	2.3	3.3	3.3	3.7	1.8	2.8	3.3	5-12	0	0
	OTHER	LO	3.3	3.0	3.1	2.8	3.3	2.4	3.3	2.8	2.7	3.3	3.5	1.7	2.9	3.3	6-12	0	0
		HI															3-9	0	0

RAND *MG811-T.E.10g*

Table E.10—continued

2-Digit AFSC	Deployed Location	Grade	Gather/Interpret	Respect Cultural Diffs	Apply Social Etiquette	Establish Credibility	Change Behavior	Verbal/Nonverbal Comm	Manage Stress	Negotiate with Others	Apply Regional Knowledge	Resolve Conflict	Self-Initiated Learning	Foreign Lang Skills	Influence Others	Establish Authority	# Respondents	Invited Sample	Population
TCN ESCORT	IRAQ/AFG	LO	3.1	3.5	2.8	3.2	2.9	2.5	3.4	2.9	3.2	3.2	3.3	2.3	3.2	3.4	**10-20**	0	0
		HI	3.0	3.3	2.6	3.0	2.7	2.0	3.0	2.7	3.3	2.6	3.6	1.6	2.6	3.6	**9-19**	0	0
	OTHER	LO	3.0	3.4	3.3	3.3	3.1	2.6	3.1	2.8	3.1	3.0	3.3	2.1	2.8	3.0	**16-38**	0	0
		HI	2.7	3.4	3.1	2.9	2.9	2.2	2.9	2.4	3.2	2.6	3.3	1.8	2.5	2.8	**11-20**	0	0

RAND *MG811-T.E.10h*

APPENDIX F

Regression Results

To test for significant differences across AFSCs, grades, and deployment locations, we conducted several regression analyses. In the first regression (see Table F.1), we computed an average score across all 14 behavior category scales to produce a single average "overall cross-cultural importance score" for the dependent variable. For the categories of applying appropriate social etiquette, self-initiated learning, and changing behavior to fit cultural context, only the shorter scale (rather than both scales) was used in calculating the overall cross-cultural importance score. For the remaining tables in this appendix, F.2 through F.18, the dependent variables were the importance ratings on each of the original 14 scales and the three shorter ("SHORT") scales.

All regressions were weighted to reflect the deployed population. Predictor variables were deployed AFSC, grade, deployment location, and deployed mission. Some airmen reported a current AFSC that differed from their deployed AFSC. Regressions reported here include those whose deployed AFSC and current AFSC did not match. We also ran a second regression that included only airmen whose current and deployed AFSCs matched. The following results were significant in one regression but not the other: 1S was significant only in the unmatched sample; 2P and 2E were significant in the matched sample. However, none of these AFSCs had large coefficients in either regression.

All R-squared values are significant at $p<.0001$.

Table F.1
Regression Results for AFSC, Grade, Deployment Location, and Deployment Location Predicting Overall Cross-Cultural Score

	Unstandardized Coefficient	Standard Error	*p*
R-squared = .1547			
Constant	2.57	.10	.00
Grade (1=low, 2=high)	–.08	.05	.07
Deployment location (1=IRAQ/AFG, 2=Other)	.08	.05	.11
Deployed mission (Flying Ops = comparison group)			
CAOC or AOC	–.13	.16	.41
Base support	.10	.09	.25
Medical	.13	.08	.11
Other ops	.35	.09	.00
Staff above Wing level	.56	.10	.00
Unknown	.21	.25	.39
AFSC (Pilot = comparison group)			
Special Investigations (OSI) (7SXXX)	1.25	.14	.00
Security Forces (31XX)	1.11	.13	.00
Special Investigations (OSI) (71XX)	1.08	.17	.00
Support Commander (30XX)	1.04	.20	.00
Contracting (64XX)	1.03	.13	.00
Contracting (6CXXX)	.89	.13	.00
Public Affairs (3NXXX)	.78	.14	.00
Security Forces (Military Police) (3PXXX)	.78	.15	.00
Surgery (45XX)	.77	.18	.00
Unallotted Airman (9UXXX)	.74	.24	.00
Public Affairs (35XX)	.73	.16	.00
Civil Engineer (32XX)	.70	.14	.00
Cardiopulmonary Laboratory (4HXXX)	.64	.25	.01
TCN Escort	.61	.16	.00
Chaplain (52XX)	.60	.14	.00
International Affairs/FAO (16XX)	.59	.17	.00
Visual Information (3VXXX)	.58	.15	.00
Aircrew Protection (1TXXX)	.56	.12	.00
Services (34XX)	.53	.14	.00
Services (3MXXX)	.52	.13	.00
Postal Specialist (8MXXX)	.52	.14	.00
No AFSC Provided	.51	.20	.01
Intelligence (1NXXX)	.50	.12	.00
Financial (6FXXX)	.49	.13	.00
Nurse (46XX)	.48	.12	.00
Physician (44XX)	.48	.15	.00
Group Superintendent (9GXXX)	.47	.21	.02

Table F.1—continued

	Unstandardized Coefficient	Standard Error	*p*
Space, Missile, and Command & Control (13XX)	.46	.13	.00
Fuels (2FXXX)	.45	.13	.00
Finance (65XX)	.45	.19	.02
Medical Service (4NXXX)	.44	.13	.00
Medical (4AXXX)	.42	.13	.00
Manpower—Personnel (37XX)	.40	.15	.01
Supply (2SXXX)	.37	.13	.01
First Sergeant (8FXXX)	.37	.14	.01
Mission Support (3SXXX)	.36	.13	.01
Chaplain Assistant (5RXXX)	.36	.14	.01
Biomedical Specialists (43XX)	.35	.18	.05
Law (51XX)	.35	.16	.03
Maintenance Management Systems (2RXXX)	.34	.15	.03
Safety (1SXXX)	.32	.15	.03
Information Management (3AXXX)	.32	.12	.01
Civil Engineering (3EXXX)	.32	.14	.02
Acquisition (63XX)	.32	.24	.17
Aerospace Medicine (48XX)	.31	.15	.05
Medical Laboratory (4TXXX)	.31	.24	.20
Intelligence (14XX)	.29	.12	.02
Weather (15XX)	.28	.18	.12
Health Services Administrator (41XX)	.27	.16	.10
Biomedical Clinician (42XX)	.27	.16	.10
Command & Control Systems Operations (1CXXX)	.25	.11	.03
Transportation & Vehicle Maintenance (2TXXX)	.25	.13	.06
Paralegal (5JXXX)	.24	.16	.12
Aircraft Maintenance (21XX)	.22	.12	.07
Aircrew Operations (1AXXX)	.20	.13	.11
Scientific/Research (61XX)	.20	.19	.31
Precision Measurement Equipment Lab (2PXXX)	.19	.17	.26
Comm-Information Systems (33XX)	.19	.13	.15
Munitions & Weapons (2WXXX)	.17	.12	.15
Manned Aerospace Maintenance (2AXXX)	.15	.12	.21
Communications & Computer Systems (3CXXX)	.15	.13	.27
Communications & Electronics (2EXXX)	.13	.12	.30
Navigator (12XX)	.12	.12	.32
Mental Health Services (4CXXX)	.11	.24	.64
Logistics Plans (2GXXX)	.05	.14	.71
Weather (1WXXX)	.02	.13	.89

Table F.2
Regression Results for AFSC, Grade, Deployment Location, and Deployed Mission Predicting Importance Ratings for Verbal and Nonverbal Communication

Verbal and Nonverbal Communication	Unstandardized Coefficient	Standard Error	*p*
R squared = .15			
Constant	2.52	.11	.00
Grade (1=low, 2=high)	–.04	.05	.49
Deployment location (1=IRAQ/AFG, 2=Other)	.12	.05	.02
Deployed mission (Flying Ops = comparison group)			
Base support	.14	.08	.10
CAOC or AOC	–.18	.16	.28
Medical	.25	.10	.02
Other ops	.44	.09	.00
Staff above Wing level	.52	.10	.00
Unknown	.39	.19	.04
AFSC (Pilot = comparison group)			
Special Investigations (OSI) (AFSC 71XX)	1.31	.18	.00
Acquisition (63XX)	.59	.27	.03
Aerospace Medicine (48XX)	.48	.18	.01
Aircraft Maintenance (21XX)	.28	.14	.04
Aircrew Operations (1AXXX)	.22	.15	.15
Aircrew Protection (1TXXX)	.57	.14	.00
Biomedical Clinician (42XX)	.44	.20	.03
Biomedical Specialists (43XX)	.39	.20	.05
Cardiopulmonary Laboratory (4HXXX)	.62	.32	.05
Chaplain (52XX)	.71	.16	.00
Chaplain Assistant (5RXXX)	.34	.18	.05
Civil Engineer (32XX)	1.03	.20	.00
Civil Engineering (3EXXX)	.37	.15	.01
Comm-Information Systems (33XX)	.23	.16	.15
Command & Control Systems Operations (1CXXX)	.33	.14	.01
Communications & Computer Systems (3CXXX)	.12	.15	.44
Communications & Electronics (2EXXX)	.16	.14	.25
Contracting (64XX)	1.25	.15	.00
Contracting (6CXXX)	1.07	.15	.00
Finance (65XX)	.51	.25	.04
Financial (6FXXX)	.48	.17	.01
First Sergeant (8FXXX)	.45	.16	.00
Fuels (2FXXX)	.53	.15	.00
Group Superintendent (9GXXX)	.41	.24	.09
Health Services Administrator (41XX)	.38	.20	.06
Information Management (3AXXX)	.30	.14	.03
Intelligence (14XX)	.21	.14	.14

Table F.2—continued

Verbal and Nonverbal Communication	Unstandardized Coefficient	Standard Error	*p*
Intelligence (1NXXX)	.59	.14	.00
International Affairs/FAO (16XX)	.69	.20	.00
Law (51XX)	.26	.20	.19
Logistics Plans (2GXXX)	.10	.16	.52
Maintenance Management Systems (2RXXX)	.34	.19	.08
Manned Aerospace Maintenance (2AXXX)	.18	.15	.21
Manpower—Personnel (37XX)	.50	.17	.00
Medical (4AXXX)	.50	.17	.00
Medical Laboratory (4TXXX)	.31	.29	.29
Medical Service (4NXXX)	.52	.16	.00
Mental Health Services (4CXXX)	.09	.32	.79
Mission Support (3SXXX)	.42	.15	.01
Munitions & Weapons (2WXXX)	.23	.14	.10
Navigator (12XX)	.22	.14	.12
No AFSC Provided	.65	.22	.00
Nurse (46XX)	.57	.14	.00
Paralegal (5JXXX)	.20	.20	.31
Physician (44XX)	.54	.19	.00
Postal Specialist (8MXXX)	.57	.16	.00
Precision Measurement Equipment Lab (2PXXX)	.20	.18	.27
Public Affairs (35XX)	.92	.19	.00
Public Affairs (3NXXX)	.90	.16	.00
Safety (1SXXX)	.37	.17	.03
Scientific/Research (61XX)	.04	.26	.87
Security Forces (Military Police) (3PXXX)	.83	.16	.00
Security Forces (31XX)	1.26	.15	.00
Services (34XX)	.64	.17	.00
Services (3MXXX)	.55	.14	.00
Space, Missile, and Command & Control (13XX)	.51	.15	.00
Special Investigations (OSI) (7SXXX)	1.51	.16	.00
Supply (2SXXX)	.31	.15	.04
Support Commander (30XX)	1.10	.21	.00
Surgery (45XX)	.95	.21	.00
TCN Escort	.85	.16	.00
Transportation & Vehicle Maintenance (2TXXX)	.30	.15	.04
Unallotted Airman (9UXXX)	1.00	.27	.00
Visual Information (3VXXX)	.74	.19	.00
Weather (15XX)	.42	.21	.05
Weather (1WXXX)	.03	.15	.85

Table F.3
Regression Results for AFSC, Grade, Deployment Location, and Deployed Mission Predicting Importance Ratings for Applying Appropriate Social Etiquette

Applying Appropriate Social Etiquette	Unstandardized Coefficient	Standard Error	*p*
R-squared = .12			
Constant	2.77	.11	.00
Grade (1=low, 2=high)	–.12	.05	.02
Deployment location (1=IRAQ/AFG, 2=Other)	.25	.05	.00
Deployed mission (Flying Ops = comparison group)			
Base support	.14	.09	.12
CAOC or AOC	.07	.17	.69
Medical	.12	.09	.20
Other ops	.35	.09	.00
Staff above Wing level	.66	.10	.00
Unknown	.31	.20	.11
AFSC (Pilot = comparison group)			
Special Investigations (OSI) (AFSC 71XX)	1.28	.18	.00
Acquisition (63XX)	.52	.27	.05
Aerospace Medicine (48XX)	.60	.18	.00
Aircraft Maintenance (21XX)	.18	.14	.21
Aircrew Operations (1AXXX)	.19	.15	.19
Aircrew Protection (1TXXX)	.59	.15	.00
Biomedical Clinician (42XX)	.51	.19	.01
Biomedical Specialists (43XX)	.51	.21	.01
Cardiopulmonary Laboratory (4HXXX)	.68	.31	.03
Chaplain (52XX)	.76	.16	.00
Chaplain Assistant (5RXXX)	.53	.18	.00
Civil Engineer (32XX)	.60	.16	.00
Civil Engineering (3EXXX)	.27	.16	.08
Comm-Information Systems (33XX)	.22	.16	.17
Command & Control Systems Operations (1CXXX)	.26	.13	.05
Communications & Computer Systems (3CXXX)	.17	.15	.28
Communications & Electronics (2EXXX)	.14	.14	.33
Contracting (64XX)	1.17	.14	.00
Contracting (6CXXX)	.98	.15	.00
Finance (65XX)	.61	.21	.00
Financial (6FXXX)	.69	.15	.00
First Sergeant (8FXXX)	.41	.18	.03
Fuels (2FXXX)	.38	.16	.02
Group Superintendent (9GXXX)	.56	.24	.02
Health Services Administrator (41XX)	.40	.18	.03
Information Management (3AXXX)	.37	.14	.01
Intelligence (14XX)	.37	.14	.01

Table F.3—continued

Applying Appropriate Social Etiquette	Unstandardized Coefficient	Standard Error	*p*
Intelligence (1NXXX)	.56	.14	.00
International Affairs/FAO (16XX)	.62	.19	.00
Law (51XX)	.33	.18	.07
Logistics Plans (2GXXX)	.01	.16	.93
Maintenance Management Systems (2RXXX)	.49	.17	.00
Manned Aerospace Maintenance (2AXXX)	.16	.15	.26
Manpower—Personnel (37XX)	.51	.16	.00
Medical (4AXXX)	.49	.16	.00
Medical Laboratory (4TXXX)	.50	.30	.09
Medical Service (4NXXX)	.44	.15	.00
Mental Health Services (4CXXX)	.29	.24	.22
Mission Support (3SXXX)	.30	.15	.05
Munitions & Weapons (2WXXX)	.06	.13	.63
Navigator (12XX)	.10	.15	.50
No AFSC Provided	.42	.21	.04
Nurse (46XX)	.60	.14	.00
Paralegal (5JXXX)	.55	.18	.00
Physician (44XX)	.80	.18	.00
Postal Specialist (8MXXX)	.35	.16	.03
Precision Measurement Equipment Lab (2PXXX)	.49	.20	.01
Public Affairs (35XX)	.86	.19	.00
Public Affairs (3NXXX)	.93	.15	.00
Safety (1SXXX)	.33	.21	.11
Scientific/Research (61XX)	.43	.22	.06
Security Forces (Military Police) (3PXXX)	.70	.16	.00
Security Forces (31XX)	.94	.18	.00
Services (34XX)	.88	.16	.00
Services (3MXXX)	.52	.15	.00
Space, Missile, and Command & Control (13XX)	.46	.15	.00
Special Investigations (OSI) (7SXXX)	1.43	.16	.00
Supply (2SXXX)	.28	.15	.05
Support Commander (30XX)	1.11	.24	.00
Surgery (45XX)	.87	.20	.00
TCN Escort	.56	.17	.00
Transportation & Vehicle Maintenance (2TXXX)	.22	.15	.14
Unallotted Airman (9UXXX)	.70	.25	.01
Visual Information (3VXXX)	.57	.17	.00
Weather (15XX)	.37	.22	.09
Weather (1WXXX)	.09	.16	.57

Table F.4
Regression Results for AFSC, Grade, Deployment Location, and Deployed Mission Predicting Importance Ratings for Managing Stress in an Unfamiliar Cultural Setting

Managing Stress in an Unfamiliar Cultural Setting	Unstandardized Coefficient	Standard Error	*p*
R-squared = .10			
Constant	2.55	.11	.00
Grade (1=low, 2=high)	–.14	.05	.01
Deployment location (1=IRAQ/AFG, 2=Other)	.05	.05	.35
Deployed mission (Flying Ops = comparison group)			
Base support	.15	.09	.09
CAOC or AOC	–.05	.17	.75
Medical	.13	.10	.21
Other ops	.27	.09	.00
Staff above Wing level	.51	.10	.00
Unknown	.29	.34	.39
AFSC (Pilot = comparison group)			
Special Investigations (OSI) (AFSC 71XX)	1.23	.17	.00
Acquisition (63XX)	.36	.26	.17
Aerospace Medicine (48XX)	.56	.19	.00
Aircraft Maintenance (21XX)	.26	.14	.07
Aircrew Operations (1AXXX)	.37	.15	.01
Aircrew Protection (1TXXX)	.78	.14	.00
Biomedical Clinician (42XX)	.37	.20	.07
Biomedical Specialists (43XX)	.64	.21	.00
Cardiopulmonary Laboratory (4HXXX)	1.02	.32	.00
Chaplain (52XX)	.70	.16	.00
Chaplain Assistant (5RXXX)	.62	.19	.00
Civil Engineer (32XX)	.57	.24	.02
Civil Engineering (3EXXX)	.37	.14	.01
Comm-Information Systems (33XX)	.32	.15	.03
Command & Control Systems Operations (1CXXX)	.39	.13	.00
Communications & Computer Systems (3CXXX)	.35	.15	.02
Communications & Electronics (2EXXX)	.26	.14	.06
Contracting (64XX)	.90	.14	.00
Contracting (6CXXX)	.89	.15	.00
Finance (65XX)	.69	.19	.00
Financial (6FXXX)	.59	.14	.00
First Sergeant (8FXXX)	.60	.16	.00
Fuels (2FXXX)	.46	.16	.00
Group Superintendent (9GXXX)	.57	.23	.02
Health Services Administrator (41XX)	.40	.20	.05
Information Management (3AXXX)	.47	.13	.00
Intelligence (14XX)	.42	.14	.00

Table F.4—continued

Managing Stress in an Unfamiliar Cultural Setting	Unstandardized Coefficient	Standard Error	*p*
Intelligence (1NXXX)	.61	.14	.00
International Affairs/FAO (16XX)	.59	.19	.00
Law (51XX)	.33	.18	.07
Logistics Plans (2GXXX)	.26	.16	.11
Maintenance Management Systems (2RXXX)	.63	.20	.00
Manned Aerospace Maintenance (2AXXX)	.35	.14	.01
Manpower—Personnel (37XX)	.46	.17	.01
Medical (4AXXX)	.61	.15	.00
Medical Laboratory (4TXXX)	.70	.25	.01
Medical Service (4NXXX)	.60	.15	.00
Mental Health Services (4CXXX)	.74	.28	.01
Mission Support (3SXXX)	.56	.15	.00
Munitions & Weapons (2WXXX)	.37	.14	.01
Navigator (12XX)	.23	.14	.11
No AFSC Provided	.43	.23	.06
Nurse (46XX)	.72	.14	.00
Paralegal (5JXXX)	.59	.19	.00
Physician (44XX)	.63	.17	.00
Postal Specialist (8MXXX)	.61	.18	.00
Precision Measurement Equipment Lab (2PXXX)	.43	.22	.05
Public Affairs (35XX)	.56	.22	.01
Public Affairs (3NXXX)	.93	.16	.00
Safety (1SXXX)	.38	.24	.11
Scientific/Research (61XX)	.38	.25	.13
Security Forces (Military Police) (3PXXX)	.85	.16	.00
Security Forces (31XX)	1.05	.15	.00
Services (34XX)	.67	.17	.00
Services (3MXXX)	.53	.15	.00
Space, Missile, and Command & Control (13XX)	.60	.14	.00
Special Investigations (OSI) (7SXXX)	1.17	.18	.00
Supply (2SXXX)	.50	.15	.00
Support Commander (30XX)	1.02	.21	.00
Surgery (45XX)	.97	.22	.00
TCN Escort	.72	.17	.00
Transportation & Vehicle Maintenance (2TXXX)	.37	.14	.01
Unallotted Airman (9UXXX)	.75	.25	.00
Visual Information (3VXXX)	.74	.17	.00
Weather (15XX)	.48	.20	.02
Weather (1WXXX)	.26	.15	.08

Table F.5
Regression Results for AFSC, Grade, Deployment Location, and Deployed Mission Predicting Importance Ratings for Changing Behavior to Fit the Cultural Context

Changing Behavior to Fit the Cultural Context	Unstandardized Coefficient	Standard Error	*p*
R-squared = .09			
Constant	2.49	.10	.00
Grade (1=low, 2=high)	–.08	.05	.16
Deployment location (1=IRAQ/AFG, 2=Other)	.16	.05	.00
Deployed mission (Flying Ops = comparison group)			
Base support	.13	.09	.15
CAOC or AOC	.02	.16	.92
Medical	.19	.09	.04
Other ops	.24	.09	.01
Staff above Wing level	.65	.11	.00
Unknown	.10	.23	.65
AFSC (Pilot = comparison group)			
Special Investigations (OSI) (AFSC 71XX)	1.32	.19	.00
Acquisition (63XX)	.58	.27	.03
Aerospace Medicine (48XX)	.34	.16	.03
Aircraft Maintenance (21XX)	.26	.12	.04
Aircrew Operations (1AXXX)	.23	.13	.08
Aircrew Protection (1TXXX)	.62	.12	.00
Biomedical Clinician (42XX)	.42	.18	.02
Biomedical Specialists (43XX)	.47	.22	.03
Cardiopulmonary Laboratory (4HXXX)	.77	.26	.00
Chaplain (52XX)	.68	.15	.00
Chaplain Assistant (5RXXX)	.48	.18	.01
Civil Engineer (32XX)	.45	.16	.00
Civil Engineering (3EXXX)	.39	.14	.01
Comm-Information Systems (33XX)	.21	.13	.12
Command & Control Systems Operations (1CXXX)	.31	.12	.01
Communications & Computer Systems (3CXXX)	.23	.14	.09
Communications & Electronics (2EXXX)	.17	.13	.16
Contracting (64XX)	1.03	.14	.00
Contracting (6CXXX)	.85	.14	.00
Finance (65XX)	.63	.19	.00
Financial (6FXXX)	.60	.13	.00
First Sergeant (8FXXX)	.40	.16	.01
Fuels (2FXXX)	.41	.14	.00
Group Superintendent (9GXXX)	.66	.22	.00
Health Services Administrator (41XX)	.40	.17	.02
Information Management (3AXXX)	.44	.12	.00
Intelligence (14XX)	.43	.12	.00

Table F.5—continued

Changing Behavior to Fit the Cultural Context	Unstandardized Coefficient	Standard Error	*p*
Intelligence (1NXXX)	.59	.12	.00
International Affairs/FAO (16XX)	.61	.17	.00
Law (51XX)	.34	.18	.06
Logistics Plans (2GXXX)	.14	.14	.34
Maintenance Management Systems (2RXXX)	.50	.15	.00
Manned Aerospace Maintenance (2AXXX)	.27	.14	.05
Manpower—Personnel (37XX)	.46	.16	.00
Medical (4AXXX)	.38	.14	.01
Medical Laboratory (4TXXX)	.32	.26	.21
Medical Service (4NXXX)	.37	.14	.01
Mental Health Services (4CXXX)	.07	.27	.80
Mission Support (3SXXX)	.42	.14	.00
Munitions & Weapons (2WXXX)	.19	.11	.09
Navigator (12XX)	.15	.13	.24
No AFSC Provided	.49	.19	.01
Nurse (46XX)	.47	.12	.00
Paralegal (5JXXX)	.43	.17	.01
Physician (44XX)	.50	.17	.00
Postal Specialist (8MXXX)	.55	.17	.00
Precision Measurement Equipment Lab (2PXXX)	.54	.19	.00
Public Affairs (35XX)	.90	.18	.00
Public Affairs (3NXXX)	.84	.16	.00
Safety (1SXXX)	.42	.19	.03
Scientific/Research (61XX)	.38	.21	.07
Security Forces (Military Police) (3PXXX)	.68	.16	.00
Security Forces (31XX)	.99	.14	.00
Services (34XX)	.57	.14	.00
Services (3MXXX)	.50	.13	.00
Space, Missile, and Command & Control (13XX)	.45	.13	.00
Special Investigations (OSI) (7SXXX)	1.39	.16	.00
Supply (2SXXX)	.42	.14	.00
Support Commander (30XX)	1.07	.21	.00
Surgery (45XX)	.67	.23	.00
TCN Escort	.48	.17	.01
Transportation & Vehicle Maintenance (2TXXX)	.29	.13	.03
Unallotted Airman (9UXXX)	.49	.24	.04
Visual Information (3VXXX)	.52	.16	.00
Weather (15XX)	.33	.21	.12
Weather (1WXXX)	.06	.14	.66

Table F.6
Regression Results for AFSC, Grade, Deployment Location, and Deployed Mission Predicting Importance Ratings for Gathering and Interpreting Observed Information

Gathering and Interpreting Observed Information	Unstandardized Coefficient	Standard Error	*p*
R-squared = .12			
Constant	2.82	.11	.00
Grade (1=low, 2=high)	–.13	.05	.01
Deployment location (1=IRAQ/AFG, 2=Other)	.14	.05	.01
Deployed mission (Flying Ops = comparison group)			
Base support	.11	.09	.23
CAOC or AOC	–.05	.19	.78
Medical	.12	.10	.20
Other ops	.38	.10	.00
Staff above Wing level	.57	.11	.00
Unknown	.49	.32	.13
AFSC (Pilot = comparison group)			
Special Investigations (OSI) (AFSC 71XX)	.95	.18	.00
Acquisition (63XX)	.62	.28	.03
Aerospace Medicine (48XX)	.34	.18	.07
Aircraft Maintenance (21XX)	.29	.14	.03
Aircrew Operations (1AXXX)	.26	.15	.08
Aircrew Protection (1TXXX)	.64	.14	.00
Biomedical Clinician (42XX)	.45	.21	.03
Biomedical Specialists (43XX)	.42	.21	.05
Cardiopulmonary Laboratory (4HXXX)	.65	.32	.04
Chaplain (52XX)	.65	.16	.00
Chaplain Assistant (5RXXX)	.34	.22	.13
Civil Engineer (32XX)	.89	.15	.00
Civil Engineering (3EXXX)	.39	.15	.01
Comm-Information Systems (33XX)	.13	.15	.37
Command & Control Systems Operations (1CXXX)	.26	.13	.05
Communications & Computer Systems (3CXXX)	.23	.15	.12
Communications & Electronics (2EXXX)	.23	.14	.11
Contracting (64XX)	1.03	.15	.00
Contracting (6CXXX)	.87	.14	.00
Finance (65XX)	.56	.20	.01
Financial (6FXXX)	.70	.15	.00
First Sergeant (8FXXX)	.47	.17	.01
Fuels (2FXXX)	.53	.15	.00
Group Superintendent (9GXXX)	.59	.24	.01
Health Services Administrator (41XX)	.34	.19	.07
Information Management (3AXXX)	.42	.13	.00
Intelligence (14XX)	.37	.14	.01

Table F.6—continued

Gathering and Interpreting Observed Information	Unstandardized Coefficient	Standard Error	*p*
Intelligence (1NXXX)	.51	.14	.00
International Affairs/FAO (16XX)	.65	.19	.00
Law (51XX)	.27	.18	.15
Logistics Plans (2GXXX)	.18	.16	.28
Maintenance Management Systems (2RXXX)	.56	.18	.00
Manned Aerospace Maintenance (2AXXX)	.30	.14	.03
Manpower—Personnel (37XX)	.49	.16	.00
Medical (4AXXX)	.52	.14	.00
Medical Laboratory (4TXXX)	.52	.27	.05
Medical Service (4NXXX)	.54	.14	.00
Mental Health Services (4CXXX)	.42	.29	.15
Mission Support (3SXXX)	.46	.15	.00
Munitions & Weapons (2WXXX)	.21	.13	.12
Navigator (12XX)	.14	.14	.34
No AFSC Provided	.72	.22	.00
Nurse (46XX)	.53	.14	.00
Paralegal (5JXXX)	.36	.18	.04
Physician (44XX)	.53	.17	.00
Postal Specialist (8MXXX)	.65	.17	.00
Precision Measurement Equipment Lab (2PXXX)	.44	.21	.03
Public Affairs (35XX)	.73	.19	.00
Public Affairs (3NXXX)	.90	.15	.00
Safety (1SXXX)	.49	.23	.03
Scientific/Research (61XX)	.23	.26	.37
Security Forces (Military Police) (3PXXX)	.84	.16	.00
Security Forces (31XX)	1.19	.14	.00
Services (34XX)	.64	.15	.00
Services (3MXXX)	.62	.14	.00
Space, Missile, and Command & Control (13XX)	.46	.14	.00
Special Investigations (OSI) (7SXXX)	1.31	.15	.00
Supply (2SXXX)	.44	.14	.00
Support Commander (30XX)	1.07	.25	.00
Surgery (45XX)	.86	.20	.00
TCN Escort	.68	.21	.00
Transportation & Vehicle Maintenance (2TXXX)	.36	.15	.02
Unallotted Airman (9UXXX)	.91	.26	.00
Visual Information (3VXXX)	.71	.16	.00
Weather (15XX)	.40	.20	.04
Weather (1WXXX)	.06	.15	.71

Table F.7
Regression Results for AFSC, Grade, Deployment Location, and Deployed Mission Predicting Importance Ratings for Applying Regional Knowledge

Applying Regional Knowledge	Unstandardized Coefficient	Standard Error	*p*
R-squared = .11			
Constant	2.78	.11	.00
Grade (1=low, 2=high)	–.06	.06	.27
Deployment location (1=IRAQ/AFG, 2=Other)	–.02	.05	.75
Deployed mission (Flying Ops = comparison group)			
Base support	.04	.09	.64
CAOC or AOC	–.19	.15	.20
Medical	.13	.09	.18
Other ops	.30	.08	.00
Staff above Wing level	.59	.10	.00
Unknown	.12	.42	.77
AFSC (Pilot = comparison group)			
Special Investigations (OSI) (AFSC 71XX)	.78	.36	.03
Acquisition (63XX)	.36	.26	.16
Aerospace Medicine (48XX)	.20	.17	.26
Aircraft Maintenance (21XX)	.06	.14	.67
Aircrew Operations (1AXXX)	–.02	.15	.91
Aircrew Protection (1TXXX)	.34	.15	.03
Biomedical Clinician (42XX)	.19	.19	.33
Biomedical Specialists (43XX)	.09	.18	.62
Cardiopulmonary Laboratory (4HXXX)	.40	.28	.15
Chaplain (52XX)	.72	.15	.00
Chaplain Assistant (5RXXX)	.21	.20	.30
Civil Engineer (32XX)	.36	.16	.03
Civil Engineering (3EXXX)	.08	.16	.62
Comm-Information Systems (33XX)	.13	.15	.40
Command & Control Systems Operations (1CXXX)	.07	.13	.60
Communications & Computer Systems (3CXXX)	–.09	.15	.58
Communications & Electronics (2EXXX)	–.07	.15	.65
Contracting (64XX)	.79	.15	.00
Contracting (6CXXX)	.49	.15	.00
Finance (65XX)	.32	.24	.18
Financial (6FXXX)	.22	.15	.13
First Sergeant (8FXXX)	.02	.17	.90
Fuels (2FXXX)	.16	.15	.29
Group Superintendent (9GXXX)	.29	.22	.18
Health Services Administrator (41XX)	–.05	.18	.76
Information Management (3AXXX)	.09	.14	.53
Intelligence (14XX)	.54	.14	.00

Table F.7—continued

Applying Regional Knowledge	Unstandardized Coefficient	Standard Error	*p*
Intelligence (1NXXX)	.46	.14	.00
International Affairs/FAO (16XX)	.51	.19	.01
Law (51XX)	.40	.17	.02
Logistics Plans (2GXXX)	–.15	.16	.37
Maintenance Management Systems (2RXXX)	.20	.17	.25
Manned Aerospace Maintenance (2AXXX)	–.07	.16	.66
Manpower—Personnel (37XX)	.27	.18	.13
Medical (4AXXX)	.10	.15	.49
Medical Laboratory (4TXXX)	–.04	.29	.90
Medical Service (4NXXX)	.11	.15	.47
Mental Health Services (4CXXX)	–.13	.26	.61
Mission Support (3SXXX)	.20	.16	.21
Munitions & Weapons (2WXXX)	–.12	.14	.38
Navigator (12XX)	.06	.13	.66
No AFSC Provided	.21	.23	.36
Nurse (46XX)	.24	.14	.10
Paralegal (5JXXX)	.02	.20	.93
Physician (44XX)	.31	.19	.11
Postal Specialist (8MXXX)	.22	.18	.21
Precision Measurement Equipment Lab (2PXXX)	–.11	.21	.62
Public Affairs (35XX)	.77	.21	.00
Public Affairs (3NXXX)	.76	.16	.00
Safety (1SXXX)	.11	.21	.61
Scientific/Research (61XX)	.20	.26	.44
Security Forces (Military Police) (3PXXX)	.51	.16	.00
Security Forces (31XX)	.90	.15	.00
Services (34XX)	.38	.17	.02
Services (3MXXX)	.24	.15	.11
Space, Missile, and Command & Control (13XX)	.35	.14	.01
Special Investigations (OSI) (7SXXX)	1.22	.16	.00
Supply (2SXXX)	.15	.15	.32
Support Commander (30XX)	.90	.28	.00
Surgery (45XX)	.37	.21	.08
TCN Escort	.27	.19	.16
Transportation & Vehicle Maintenance (2TXXX)	.09	.15	.53
Unallotted Airman (9UXXX)	.31	.24	.19
Visual Information (3VXXX)	.19	.17	.27
Weather (15XX)	.10	.22	.63
Weather (1WXXX)	–.16	.15	.29

Table F.8
Regression Results for AFSC, Grade, Deployment Location, and Deployed Mission Predicting Importance Ratings for Self-Initiated Learning

Self-Initiated Learning	Unstandardized Coefficient	Standard Error	*p*
R-squared = .12			
Constant	2.49	.10	.00
Grade (1=low, 2=high)	–.14	.05	.01
Deployment location (1=IRAQ/AFG, 2=Other)	.13	.05	.01
Deployed mission (Flying Ops = comparison group)			
Base support	.04	.09	.66
CAOC or AOC	–.21	.16	.21
Medical	.05	.09	.61
Other ops	.25	.09	.01
Staff above Wing level	.54	.11	.00
Unknown	.15	.19	.43
AFSC (Pilot = comparison group)			
Special Investigations (OSI) (AFSC 71XX)	.94	.15	.00
Acquisition (63XX)	.61	.32	.05
Aerospace Medicine (48XX)	.34	.16	.03
Aircraft Maintenance (21XX)	.21	.12	.09
Aircrew Operations (1AXXX)	.14	.13	.28
Aircrew Protection (1TXXX)	.63	.12	.00
Biomedical Clinician (42XX)	.44	.17	.01
Biomedical Specialists (43XX)	.46	.16	.01
Cardiopulmonary Laboratory (4HXXX)	.72	.24	.00
Chaplain (52XX)	.73	.15	.00
Chaplain Assistant (5RXXX)	.53	.18	.00
Civil Engineer (32XX)	.49	.14	.00
Civil Engineering (3EXXX)	.34	.14	.02
Comm-Information Systems (33XX)	.25	.13	.06
Command & Control Systems Operations (1CXXX)	.26	.11	.02
Communications & Computer Systems (3CXXX)	.16	.14	.25
Communications & Electronics (2EXXX)	.12	.13	.34
Contracting (64XX)	.99	.14	.00
Contracting (6CXXX)	.85	.14	.00
Finance (65XX)	.46	.24	.06
Financial (6FXXX)	.70	.14	.00
First Sergeant (8FXXX)	.34	.17	.05
Fuels (2FXXX)	.26	.13	.05
Group Superintendent (9GXXX)	.47	.21	.03
Health Services Administrator (41XX)	.52	.17	.00
Information Management (3AXXX)	.34	.12	.01
Intelligence (14XX)	.36	.13	.01

Table F.8—continued

Self-Initiated Learning	Unstandardized Coefficient	Standard Error	*p*
Intelligence (1NXXX)	.53	.12	.00
International Affairs/FAO (16XX)	.65	.18	.00
Law (51XX)	.38	.17	.03
Logistics Plans (2GXXX)	.13	.15	.37
Maintenance Management Systems (2RXXX)	.28	.16	.07
Manned Aerospace Maintenance (2AXXX)	.15	.13	.26
Manpower—Personnel (37XX)	.44	.14	.00
Medical (4AXXX)	.55	.14	.00
Medical Laboratory (4TXXX)	.55	.22	.01
Medical Service (4NXXX)	.56	.13	.00
Mental Health Services (4CXXX)	.01	.20	.98
Mission Support (3SXXX)	.46	.14	.00
Munitions & Weapons (2WXXX)	.15	.12	.20
Navigator (12XX)	.13	.13	.31
No AFSC Provided	.57	.19	.00
Nurse (46XX)	.55	.13	.00
Paralegal (5JXXX)	.46	.16	.00
Physician (44XX)	.55	.19	.00
Postal Specialist (8MXXX)	.68	.15	.00
Precision Measurement Equipment Lab (2PXXX)	.24	.16	.14
Public Affairs (35XX)	.63	.19	.00
Public Affairs (3NXXX)	.80	.15	.00
Safety (1SXXX)	.44	.16	.01
Scientific/Research (61XX)	.50	.21	.02
Security Forces (Military Police) (3PXXX)	.77	.16	.00
Security Forces (31XX)	.95	.15	.00
Services (34XX)	.59	.15	.00
Services (3MXXX)	.48	.13	.00
Space, Missile, and Command & Control (13XX)	.48	.13	.00
Special Investigations (OSI) (7SXXX)	1.13	.16	.00
Supply (2SXXX)	.37	.14	.01
Support Commander (30XX)	.88	.19	.00
Surgery (45XX)	.82	.20	.00
TCN Escort	.44	.17	.01
Transportation & Vehicle Maintenance (2TXXX)	.19	.13	.15
Unallotted Airman (9UXXX)	.62	.23	.01
Visual Information (3VXXX)	.61	.16	.00
Weather (15XX)	.35	.19	.06
Weather (1WXXX)	.17	.14	.21

Table F.9
Regression Results for AFSC, Grade, Deployment Location, and Deployed Mission Predicting Importance Ratings for Respecting Cultural Differences

Respecting Cultural Differences	Unstandardized Coefficient	Standard Error	*p*
R-squared = .10			
Constant	2.87	.11	.00
Grade (1=low, 2=high)	–.11	.05	.05
Deployment location (1=IRAQ/AFG, 2=Other)	.19	.05	.00
Deployed mission (Flying Ops = comparison group)			
Base support	.10	.09	.26
CAOC or AOC	–.04	.16	.79
Medical	–.01	.10	.96
Other ops	.27	.09	.00
Staff above Wing level	.48	.08	.00
Unknown	.13	.26	.62
AFSC (Pilot = comparison group)			
Special Investigations (OSI) (AFSC 71XX)	.87	.18	.00
Acquisition (63XX)	.52	.29	.07
Aerospace Medicine (48XX)	.57	.16	.00
Aircraft Maintenance (21XX)	.31	.14	.03
Aircrew Operations (1AXXX)	.24	.15	.12
Aircrew Protection (1TXXX)	.52	.14	.00
Biomedical Clinician (42XX)	.58	.19	.00
Biomedical Specialists (43XX)	.49	.19	.01
Cardiopulmonary Laboratory (4HXXX)	.79	.32	.01
Chaplain (52XX)	.82	.15	.00
Chaplain Assistant (5RXXX)	.62	.19	.00
Civil Engineer (32XX)	.66	.18	.00
Civil Engineering (3EXXX)	.30	.16	.06
Comm-Information Systems (33XX)	.23	.15	.14
Command & Control Systems Operations (1CXXX)	.30	.13	.02
Communications & Computer Systems (3CXXX)	.21	.15	.17
Communications & Electronics (2EXXX)	.18	.14	.19
Contracting (64XX)	1.17	.14	.00
Contracting (6CXXX)	1.00	.14	.00
Finance (65XX)	.63	.20	.00
Financial (6FXXX)	.71	.15	.00
First Sergeant (8FXXX)	.57	.17	.00
Fuels (2FXXX)	.35	.14	.02
Group Superintendent (9GXXX)	.58	.24	.02
Health Services Administrator (41XX)	.55	.19	.00
Information Management (3AXXX)	.39	.13	.00
Intelligence (14XX)	.46	.14	.00

Table F.9—continued

Respecting Cultural Differences	Unstandardized Coefficient	Standard Error	*p*
Intelligence (1NXXX)	.54	.14	.00
International Affairs/FAO (16XX)	.58	.19	.00
Law (51XX)	.39	.19	.04
Logistics Plans (2GXXX)	.04	.17	.80
Maintenance Management Systems (2RXXX)	.52	.18	.00
Manned Aerospace Maintenance (2AXXX)	.20	.15	.17
Manpower—Personnel (37XX)	.58	.15	.00
Medical (4AXXX)	.69	.15	.00
Medical Laboratory (4TXXX)	.70	.28	.01
Medical Service (4NXXX)	.60	.15	.00
Mental Health Services (4CXXX)	.71	.27	.01
Mission Support (3SXXX)	.43	.15	.01
Munitions & Weapons (2WXXX)	.06	.14	.67
Navigator (12XX)	.11	.14	.45
No AFSC Provided	.53	.22	.02
Nurse (46XX)	.75	.14	.00
Paralegal (5JXXX)	.44	.18	.01
Physician (44XX)	1.02	.16	.00
Postal Specialist (8MXXX)	.54	.16	.00
Precision Measurement Equipment Lab (2PXXX)	.41	.18	.02
Public Affairs (35XX)	.85	.18	.00
Public Affairs (3NXXX)	.89	.16	.00
Safety (1SXXX)	.31	.15	.04
Scientific/Research (61XX)	.47	.20	.02
Security Forces (Military Police) (3PXXX)	.71	.16	.00
Security Forces (31XX)	.99	.16	.00
Services (34XX)	.74	.14	.00
Services (3MXXX)	.49	.14	.00
Space, Missile, and Command & Control (13XX)	.47	.15	.00
Special Investigations (OSI) (7SXXX)	1.10	.18	.00
Supply (2SXXX)	.36	.15	.02
Support Commander (30XX)	1.15	.18	.00
Surgery (45XX)	1.07	.19	.00
TCN Escort	.50	.19	.01
Transportation & Vehicle Maintenance (2TXXX)	.21	.15	.16
Unallotted Airman (9UXXX)	.80	.26	.00
Visual Information (3VXXX)	.67	.18	.00
Weather (15XX)	.47	.18	.01
Weather (1WXXX)	.11	.16	.48

Table F.10
Regression Results for AFSC, Grade, Deployment Location, and Deployed Mission Predicting Importance Ratings for Establishing Authority

Establishing Authority	Unstandardized Coefficient	Standard Error	*p*
R-squared = .22			
Constant	2.20	.09	.00
Grade (1=low, 2=high)	–.16	.05	.00
Deployment location (1=IRAQ/AFG, 2=Other)	–.14	.05	.00
Deployed mission (Flying Ops = comparison group)			
Base support	.19	.09	.04
CAOC or AOC	–.24	.16	.15
Medical	.10	.10	.32
Other ops	.36	.08	.00
Staff above Wing level	.37	.16	.02
Unknown	–.23	.18	.19
AFSC (Pilot = comparison group)			
Special Investigations (OSI) (AFSC 71XX)	.99	.28	.00
Acquisition (63XX)	.30	.18	.10
Aerospace Medicine (48XX)	–.10	.14	.48
Aircraft Maintenance (21XX)	.15	.12	.24
Aircrew Operations (1AXXX)	.21	.12	.09
Aircrew Protection (1TXXX)	.40	.13	.00
Biomedical Clinician (42XX)	–.10	.16	.54
Biomedical Specialists (43XX)	.18	.19	.33
Cardiopulmonary Laboratory (4HXXX)	.65	.29	.03
Chaplain (52XX)	–.14	.15	.35
Chaplain Assistant (5RXXX)	.21	.20	.28
Civil Engineer (32XX)	.67	.19	.00
Civil Engineering (3EXXX)	.53	.15	.00
Comm-Information Systems (33XX)	–.01	.13	.97
Command & Control Systems Operations (1CXXX)	.31	.12	.01
Communications & Computer Systems (3CXXX)	.15	.14	.26
Communications & Electronics (2EXXX)	.19	.12	.12
Contracting (64XX)	.59	.14	.00
Contracting (6CXXX)	.70	.14	.00
Finance (65XX)	.01	.19	.96
Financial (6FXXX)	.25	.14	.06
First Sergeant (8FXXX)	.21	.14	.15
Fuels (2FXXX)	.59	.14	.00
Group Superintendent (9GXXX)	.38	.22	.09
Health Services Administrator (41XX)	.05	.18	.76
Information Management (3AXXX)	.33	.12	.01
Intelligence (14XX)	.00	.13	.98

Table F.10—continued

Establishing Authority	Unstandardized Coefficient	Standard Error	*p*
Intelligence (1NXXX)	.26	.12	.04
International Affairs/FAO (16XX)	.18	.16	.25
Law (51XX)	.34	.17	.04
Logistics Plans (2GXXX)	.08	.14	.55
Maintenance Management Systems (2RXXX)	.15	.16	.37
Manned Aerospace Maintenance (2AXXX)	.31	.12	.01
Manpower—Personnel (37XX)	.01	.14	.95
Medical (4AXXX)	.44	.14	.00
Medical Laboratory (4TXXX)	.17	.22	.45
Medical Service (4NXXX)	.50	.15	.00
Mental Health Services (4CXXX)	–.05	.26	.84
Mission Support (3SXXX)	.34	.14	.01
Munitions & Weapons (2WXXX)	.44	.12	.00
Navigator (12XX)	.04	.12	.75
No AFSC Provided	.43	.22	.06
Nurse (46XX)	.26	.13	.05
Paralegal (5JXXX)	.12	.19	.52
Physician (44XX)	.02	.16	.89
Postal Specialist (8MXXX)	.53	.18	.00
Precision Measurement Equipment Lab (2PXXX)	–.05	.19	.77
Public Affairs (35XX)	–.09	.16	.55
Public Affairs (3NXXX)	.24	.16	.13
Safety (1SXXX)	.44	.18	.02
Scientific/Research (61XX)	–.09	.22	.69
Security Forces (Military Police) (3PXXX)	1.09	.16	.00
Security Forces (31XX)	1.31	.17	.00
Services (34XX)	.09	.16	.58
Services (3MXXX)	.57	.14	.00
Space, Missile, and Command & Control (13XX)	.24	.13	.06
Special Investigations (OSI) (7SXXX)	.91	.19	.00
Supply (2SXXX)	.37	.14	.01
Support Commander (30XX)	.67	.26	.01
Surgery (45XX)	.39	.21	.07
TCN Escort	1.03	.17	.00
Transportation & Vehicle Maintenance (2TXXX)	.41	.13	.00
Unallotted Airman (9UXXX)	1.14	.25	.00
Visual Information (3VXXX)	.37	.17	.03
Weather (15XX)	–.11	.19	.56
Weather (1WXXX)	–.05	.13	.67

Table F.11
Regression Results for AFSC, Grade, Deployment Location, and Deployed Mission Predicting Importance Ratings for Influencing Others

Influencing Others	Unstandardized Coefficient	Standard Error	*p*
R-squared = .18			
Constant	2.31	.10	.00
Grade (1=low, 2=high)	–.03	.05	.53
Deployment location (1=IRAQ/AFG, 2=Other)	–.08	.05	.09
Deployed mission (Flying Ops = comparison group)			
Base support	.14	.10	.15
CAOC or AOC	–.20	.16	.21
Medical	.14	.09	.13
Other ops	.42	.09	.00
Staff above Wing level	.51	.10	.00
Unknown	–.05	.18	.77
AFSC (Pilot = comparison group)			
Special Investigations (OSI) (AFSC 71XX)	1.17	.22	.00
Acquisition (63XX)	.46	.25	.06
Aerospace Medicine (48XX)	.28	.17	.11
Aircraft Maintenance (21XX)	.26	.13	.04
Aircrew Operations (1AXXX)	.18	.13	.17
Aircrew Protection (1TXXX)	.44	.13	.00
Biomedical Clinician (42XX)	.33	.19	.08
Biomedical Specialists (43XX)	.32	.19	.09
Cardiopulmonary Laboratory (4HXXX)	.60	.22	.01
Chaplain (52XX)	.34	.16	.04
Chaplain Assistant (5RXXX)	.16	.17	.33
Civil Engineer (32XX)	.68	.13	.00
Civil Engineering (3EXXX)	.40	.15	.01
Comm-Information Systems (33XX)	.19	.14	.18
Command & Control Systems Operations (1CXXX)	.20	.12	.09
Communications & Computer Systems (3CXXX)	.08	.14	.55
Communications & Electronics (2EXXX)	.04	.13	.78
Contracting (64XX)	.91	.15	.00
Contracting (6CXXX)	.76	.14	.00
Finance (65XX)	.29	.21	.17
Financial (6FXXX)	.27	.14	.05
First Sergeant (8FXXX)	.23	.16	.15
Fuels (2FXXX)	.49	.14	.00
Group Superintendent (9GXXX)	.38	.23	.10
Health Services Administrator (41XX)	.21	.18	.23
Information Management (3AXXX)	.25	.12	.05
Intelligence (14XX)	.06	.13	.63

Table F.11—continued

Influencing Others	Unstandardized Coefficient	Standard Error	*p*
Intelligence (1NXXX)	.35	.13	.01
International Affairs/FAO (16XX)	.65	.18	.00
Law (51XX)	.34	.18	.06
Logistics Plans (2GXXX)	–.02	.15	.89
Maintenance Management Systems (2RXXX)	.15	.15	.32
Manned Aerospace Maintenance (2AXXX)	.17	.13	.18
Manpower—Personnel (37XX)	.26	.15	.08
Medical (4AXXX)	.39	.14	.01
Medical Laboratory (4TXXX)	.31	.30	.29
Medical Service (4NXXX)	.47	.14	.00
Mental Health Services (4CXXX)	.07	.25	.77
Mission Support (3SXXX)	.29	.14	.04
Munitions & Weapons (2WXXX)	.23	.12	.07
Navigator (12XX)	.13	.12	.27
No AFSC Provided	.53	.21	.01
Nurse (46XX)	.38	.13	.00
Paralegal (5JXXX)	–.02	.19	.90
Physician (44XX)	.37	.17	.03
Postal Specialist (8MXXX)	.56	.17	.00
Precision Measurement Equipment Lab (2PXXX)	–.13	.18	.49
Public Affairs (35XX)	.71	.19	.00
Public Affairs (3NXXX)	.54	.17	.00
Safety (1SXXX)	.32	.17	.06
Scientific/Research (61XX)	–.05	.20	.79
Security Forces (Military Police) (3PXXX)	.80	.16	.00
Security Forces (31XX)	1.26	.16	.00
Services (34XX)	.34	.17	.05
Services (3MXXX)	.51	.14	.00
Space, Missile, and Command & Control (13XX)	.38	.14	.01
Special Investigations (OSI) (7SXXX)	1.33	.17	.00
Supply (2SXXX)	.36	.14	.01
Support Commander (30XX)	1.08	.29	.00
Surgery (45XX)	.72	.21	.00
TCN Escort	.84	.15	.00
Transportation & Vehicle Maintenance (2TXXX)	.29	.13	.03
Unallotted Airman (9UXXX)	.85	.24	.00
Visual Information (3VXXX)	.50	.16	.00
Weather (15XX)	.06	.18	.73
Weather (1WXXX)	–.06	.14	.66

Table F.12
Regression Results for AFSC, Grade, Deployment Location, and Deployed Mission Predicting Importance Ratings for Negotiating with Others

Negotiating with Others	Unstandardized Coefficient	Standard Error	*p*
R-squared = .15			
Constant	2.51	.11	.00
Grade (1=low, 2=high)	–.09	.05	.10
Deployment location (1=IRAQ/AFG, 2=Other)	.13	.05	.01
Deployed mission (Flying Ops = comparison group)			
Base support	.11	.09	.24
CAOC or AOC	–.09	.17	.58
Medical	.03	.12	.83
Other ops	.35	.09	.00
Staff above Wing level	.55	.13	.00
Unknown	.36	.30	.24
AFSC (Pilot = comparison group)			
Special Investigations (OSI) (AFSC 71XX)	1.39	.18	.00
Acquisition (63XX)	.40	.25	.11
Aerospace Medicine (48XX)	.49	.19	.01
Aircraft Maintenance (21XX)	.22	.14	.12
Aircrew Operations (1AXXX)	.23	.14	.11
Aircrew Protection (1TXXX)	.49	.14	.00
Biomedical Clinician (42XX)	.28	.22	.21
Biomedical Specialists (43XX)	.51	.21	.02
Cardiopulmonary Laboratory (4HXXX)	.70	.29	.02
Chaplain (52XX)	.52	.16	.00
Chaplain Assistant (5RXXX)	.41	.16	.01
Civil Engineer (32XX)	.81	.18	.00
Civil Engineering (3EXXX)	.35	.16	.03
Comm-Information Systems (33XX)	.27	.15	.08
Command & Control Systems Operations (1CXXX)	.22	.13	.09
Communications & Computer Systems (3CXXX)	.16	.15	.28
Communications & Electronics (2EXXX)	.12	.13	.39
Contracting (64XX)	1.65	.14	.00
Contracting (6CXXX)	1.46	.14	.00
Finance (65XX)	.47	.25	.06
Financial (6FXXX)	.54	.15	.00
First Sergeant (8FXXX)	.31	.16	.06
Fuels (2FXXX)	.54	.15	.00
Group Superintendent (9GXXX)	.29	.24	.22
Health Services Administrator (41XX)	.32	.19	.09
Information Management (3AXXX)	.26	.13	.05
Intelligence (14XX)	.20	.14	.14

Table F.12—continued

Negotiating with Others	Unstandardized Coefficient	Standard Error	*p*
Intelligence (1NXXX)	.50	.13	.00
International Affairs/FAO (16XX)	.65	.20	.00
Law (51XX)	.44	.19	.02
Logistics Plans (2GXXX)	–.02	.16	.90
Maintenance Management Systems (2RXXX)	.25	.17	.14
Manned Aerospace Maintenance (2AXXX)	.09	.14	.52
Manpower—Personnel (37XX)	.41	.18	.02
Medical (4AXXX)	.39	.16	.01
Medical Laboratory (4TXXX)	.32	.29	.27
Medical Service (4NXXX)	.44	.16	.01
Mental Health Services (4CXXX)	.05	.35	.88
Mission Support (3SXXX)	.33	.15	.03
Munitions & Weapons (2WXXX)	.13	.13	.33
Navigator (12XX)	.09	.13	.49
No AFSC Provided	.65	.23	.01
Nurse (46XX)	.45	.15	.00
Paralegal (5JXXX)	.01	.18	.98
Physician (44XX)	.54	.19	.00
Postal Specialist (8MXXX)	.30	.17	.08
Precision Measurement Equipment Lab (2PXXX)	.21	.20	.30
Public Affairs (35XX)	.78	.20	.00
Public Affairs (3NXXX)	.81	.16	.00
Safety (1SXXX)	.26	.18	.14
Scientific/Research (61XX)	.10	.26	.70
Security Forces (Military Police) (3PXXX)	.76	.16	.00
Security Forces (31XX)	1.17	.15	.00
Services (34XX)	.76	.17	.00
Services (3MXXX)	.55	.14	.00
Space, Missile, and Command & Control (13XX)	.52	.15	.00
Special Investigations (OSI) (7SXXX)	1.51	.15	.00
Supply (2SXXX)	.34	.14	.02
Support Commander (30XX)	1.15	.25	.00
Surgery (45XX)	.70	.23	.00
TCN Escort	.50	.17	.00
Transportation & Vehicle Maintenance (2TXXX)	.24	.14	.09
Unallotted Airman (9UXXX)	.42	.25	.10
Visual Information (3VXXX)	.59	.17	.00
Weather (15XX)	.30	.22	.18
Weather (1WXXX)	–.02	.16	.90

Table F.13
Regression Results for AFSC, Grade, Deployment Location, and Deployed Mission Predicting Importance Ratings for Establishing Credibility, Trust, and Respect

Establishing Credibility, Trust, and Respect	Unstandardized Coefficient	Standard Error	*p*
R-squared = .14			
Constant	2.63	.11	.00
Grade (1=low, 2=high)	–.05	.05	.34
Deployment location (1=IRAQ/AFG, 2=Other)	.06	.05	.23
Deployed mission (Flying Ops = comparison group)			
Base support	.11	.09	.24
CAOC or AOC	–.23	.17	.19
Medical	.18	.10	.08
Other ops	.41	.09	.00
Staff above Wing level	.62	.10	.00
Unknown	.26	.27	.34
AFSC (Pilot = comparison group)			
Special Investigations (OSI) (AFSC 71XX)	1.11	.24	.00
Acquisition (63XX)	.35	.26	.18
Aerospace Medicine (48XX)	.46	.19	.02
Aircraft Maintenance (21XX)	.23	.14	.12
Aircrew Operations (1AXXX)	.21	.15	.18
Aircrew Protection (1TXXX)	.52	.14	.00
Biomedical Clinician (42XX)	.36	.20	.07
Biomedical Specialists (43XX)	.37	.21	.08
Cardiopulmonary Laboratory (4HXXX)	.63	.33	.06
Chaplain (52XX)	.62	.17	.00
Chaplain Assistant (5RXXX)	.37	.16	.02
Civil Engineer (32XX)	.75	.16	.00
Civil Engineering (3EXXX)	.24	.16	.13
Comm-Information Systems (33XX)	.16	.16	.31
Command & Control Systems Operations (1CXXX)	.18	.13	.18
Communications & Computer Systems (3CXXX)	.09	.15	.55
Communications & Electronics (2EXXX)	.12	.14	.39
Contracting (64XX)	1.24	.14	.00
Contracting (6CXXX)	1.03	.15	.00
Finance (65XX)	.50	.22	.02
Financial (6FXXX)	.44	.15	.00
First Sergeant (8FXXX)	.36	.18	.05
Fuels (2FXXX)	.37	.15	.01
Group Superintendent (9GXXX)	.43	.24	.08
Health Services Administrator (41XX)	.22	.18	.23
Information Management (3AXXX)	.22	.14	.10
Intelligence (14XX)	.19	.15	.21

Table F.13—continued

Establishing Credibility, Trust, and Respect	Unstandardized Coefficient	Standard Error	*p*
Intelligence (1NXXX)	.41	.14	.00
International Affairs/FAO (16XX)	.64	.20	.00
Law (51XX)	.40	.20	.05
Logistics Plans (2GXXX)	–.03	.16	.85
Maintenance Management Systems (2RXXX)	.21	.19	.25
Manned Aerospace Maintenance (2AXXX)	.15	.14	.30
Manpower—Personnel (37XX)	.36	.17	.04
Medical (4AXXX)	.36	.16	.02
Medical Laboratory (4TXXX)	.40	.28	.15
Medical Service (4NXXX)	.42	.15	.01
Mental Health Services (4CXXX)	–.05	.29	.88
Mission Support (3SXXX)	.26	.15	.09
Munitions & Weapons (2WXXX)	.11	.14	.40
Navigator (12XX)	.12	.14	.40
No AFSC Provided	.49	.22	.03
Nurse (46XX)	.45	.14	.00
Paralegal (5JXXX)	.13	.18	.48
Physician (44XX)	.64	.18	.00
Postal Specialist (8MXXX)	.40	.16	.01
Precision Measurement Equipment Lab (2PXXX)	.05	.19	.77
Public Affairs (35XX)	.75	.17	.00
Public Affairs (3NXXX)	.78	.16	.00
Safety (1SXXX)	.18	.17	.27
Scientific/Research (61XX)	.25	.21	.23
Security Forces (Military Police) (3PXXX)	.75	.17	.00
Security Forces (31XX)	.84	.27	.00
Services (34XX)	.60	.17	.00
Services (3MXXX)	.52	.14	.00
Space, Missile, and Command & Control (13XX)	.53	.15	.00
Special Investigations (OSI) (7SXXX)	1.30	.17	.00
Supply (2SXXX)	.35	.15	.02
Support Commander (30XX)	1.27	.23	.00
Surgery (45XX)	.95	.20	.00
TCN Escort	.56	.21	.01
Transportation & Vehicle Maintenance (2TXXX)	.25	.15	.09
Unallotted Airman (9UXXX)	.69	.28	.01
Visual Information (3VXXX)	.59	.19	.00
Weather (15XX)	.21	.21	.31
Weather (1WXXX)	.01	.16	.97

Table F.14
Regression Results for AFSC, Grade, Deployment Location, and Deployed Mission Predicting Importance Ratings for Resolving Conflict

Resolving Conflict	Unstandardized Coefficient	Standard Error	*p*
R-squared = .14			
Constant	2.39	.10	.00
Grade (1=low, 2=high)	–.07	.05	.18
Deployment location (1=IRAQ/AFG, 2=Other)	–.05	.05	.32
Deployed mission (Flying Ops = comparison group)			
Base support	.16	.09	.08
CAOC or AOC	–.17	.15	.27
Medical	.13	.09	.18
Other ops	.36	.09	.00
Staff above Wing level	.52	.13	.00
Unknown	.28	.29	.34
AFSC (Pilot = comparison group)			
Special Investigations (OSI) (AFSC 71XX)	1.04	.22	.00
Acquisition (63XX)	.38	.30	.21
Aerospace Medicine (48XX)	–.01	.18	.97
Aircraft Maintenance (21XX)	.19	.13	.15
Aircrew Operations (1AXXX)	.17	.14	.20
Aircrew Protection (1TXXX)	.50	.13	.00
Biomedical Clinician (42XX)	.05	.19	.78
Biomedical Specialists (43XX)	.28	.19	.14
Cardiopulmonary Laboratory (4HXXX)	.51	.23	.03
Chaplain (52XX)	.32	.16	.05
Chaplain Assistant (5RXXX)	.16	.20	.42
Civil Engineer (32XX)	.73	.21	.00
Civil Engineering (3EXXX)	.26	.15	.08
Comm-Information Systems (33XX)	.10	.15	.47
Command & Control Systems Operations (1CXXX)	.28	.13	.03
Communications & Computer Systems (3CXXX)	.11	.14	.45
Communications & Electronics (2EXXX)	.11	.13	.40
Contracting (64XX)	.83	.17	.00
Contracting (6CXXX)	.71	.15	.00
Finance (65XX)	.26	.23	.26
Financial (6FXXX)	.25	.14	.09
First Sergeant (8FXXX)	.36	.18	.05
Fuels (2FXXX)	.41	.15	.01
Group Superintendent (9GXXX)	.48	.22	.03
Health Services Administrator (41XX)	.00	.20	.98
Information Management (3AXXX)	.27	.13	.04
Intelligence (14XX)	.06	.13	.63

Table F.14—continued

Resolving Conflict	Unstandardized Coefficient	Standard Error	*p*
Intelligence (1NXXX)	.40	.14	.00
International Affairs/FAO (16XX)	.53	.19	.01
Law (51XX)	.38	.18	.04
Logistics Plans (2GXXX)	.02	.15	.90
Maintenance Management Systems (2RXXX)	.30	.17	.08
Manned Aerospace Maintenance (2AXXX)	.27	.14	.05
Manpower—Personnel (37XX)	.25	.16	.12
Medical (4AXXX)	.31	.15	.03
Medical Laboratory (4TXXX)	.23	.28	.40
Medical Service (4NXXX)	.41	.14	.01
Mental Health Services (4CXXX)	.18	.30	.55
Mission Support (3SXXX)	.35	.15	.02
Munitions & Weapons (2WXXX)	.26	.13	.04
Navigator (12XX)	.03	.13	.82
No AFSC Provided	.45	.25	.08
Nurse (46XX)	.35	.14	.01
Paralegal (5JXXX)	.05	.20	.78
Physician (44XX)	.18	.19	.34
Postal Specialist (8MXXX)	.69	.17	.00
Precision Measurement Equipment Lab (2PXXX)	.20	.23	.38
Public Affairs (35XX)	.72	.21	.00
Public Affairs (3NXXX)	.42	.19	.03
Safety (1SXXX)	.43	.17	.01
Scientific/Research (61XX)	–.30	.23	.19
Security Forces (Military Police) (3PXXX)	.87	.16	.00
Security Forces (31XX)	1.30	.14	.00
Services (34XX)	.19	.16	.22
Services (3MXXX)	.54	.15	.00
Space, Missile, and Command & Control (13XX)	.37	.14	.01
Special Investigations (OSI) (7SXXX)	1.08	.19	.00
Supply (2SXXX)	.46	.14	.00
Support Commander (30XX)	1.02	.31	.00
Surgery (45XX)	.48	.23	.04
TCN Escort	.71	.17	.00
Transportation & Vehicle Maintenance (2TXXX)	.33	.14	.02
Unallotted Airman (9UXXX)	.83	.25	.00
Visual Information (3VXXX)	.45	.17	.01
Weather (15XX)	–.01	.23	.96
Weather (1WXXX)	–.09	.14	.49

Table F.15
Regression Results for AFSC, Grade, Deployment Location, and Deployed Mission Predicting Importance Ratings for Foreign Language Skills

Foreign Language Skills	Unstandardized Coefficient	Standard Error	*p*
R-squared = .10			
Constant	2.43	.11	.00
Grade (1=low, 2=high)	.00	.06	.98
Deployment location (1=IRAQ/AFG, 2=Other)	.06	.06	.30
Deployed mission (Flying Ops = comparison group)			
Base support	.04	.09	.67
CAOC or AOC	–.11	.15	.46
Medical	.22	.11	.04
Other ops	.37	.10	.00
Staff above Wing level	.59	.14	.00
Unknown	.34	.39	.38
AFSC (Pilot = comparison group)			
Special Investigations (OSI) (AFSC 71XX)	.51	.32	.12
Acquisition (63XX)	.22	.29	.45
Aerospace Medicine (48XX)	–.02	.17	.91
Aircraft Maintenance (21XX)	.13	.14	.35
Aircrew Operations (1AXXX)	.25	.16	.10
Aircrew Protection (1TXXX)	.48	.13	.00
Biomedical Clinician (42XX)	–.13	.19	.50
Biomedical Specialists (43XX)	.06	.18	.74
Cardiopulmonary Laboratory (4HXXX)	.40	.25	.11
Chaplain (52XX)	.54	.18	.00
Chaplain Assistant (5RXXX)	.26	.17	.14
Civil Engineer (32XX)	.43	.17	.01
Civil Engineering (3EXXX)	.19	.16	.23
Comm-Information Systems (33XX)	.13	.16	.41
Command & Control Systems Operations (1CXXX)	.02	.13	.90
Communications & Computer Systems (3CXXX)	.01	.15	.95
Communications & Electronics (2EXXX)	.01	.14	.94
Contracting (64XX)	.71	.17	.00
Contracting (6CXXX)	.35	.16	.03
Finance (65XX)	.04	.20	.85
Financial (6FXXX)	.28	.15	.06
First Sergeant (8FXXX)	.26	.17	.13
Fuels (2FXXX)	.31	.15	.05
Group Superintendent (9GXXX)	.19	.24	.42
Health Services Administrator (41XX)	.10	.20	.62
Information Management (3AXXX)	.19	.14	.16
Intelligence (14XX)	.22	.15	.15

Table F.15—continued

Foreign Language Skills	Unstandardized Coefficient	Standard Error	*p*
Intelligence (1NXXX)	.25	.14	.07
International Affairs/FAO (16XX)	.36	.20	.07
Law (51XX)	.10	.20	.62
Logistics Plans (2GXXX)	–.07	.17	.66
Maintenance Management Systems (2RXXX)	.16	.17	.33
Manned Aerospace Maintenance (2AXXX)	–.07	.16	.67
Manpower—Personnel (37XX)	.40	.19	.04
Medical (4AXXX)	.18	.16	.26
Medical Laboratory (4TXXX)	.03	.25	.91
Medical Service (4NXXX)	.22	.15	.16
Mental Health Services (4CXXX)	–.69	.26	.01
Mission Support (3SXXX)	.28	.15	.06
Munitions & Weapons (2WXXX)	.04	.14	.76
Navigator (12XX)	.17	.14	.23
No AFSC Provided	.40	.23	.08
Nurse (46XX)	.34	.15	.02
Paralegal (5JXXX)	.05	.18	.77
Physician (44XX)	.13	.21	.54
Postal Specialist (8MXXX)	.60	.20	.00
Precision Measurement Equipment Lab (2PXXX)	.18	.23	.43
Public Affairs (35XX)	.82	.26	.00
Public Affairs (3NXXX)	.51	.20	.01
Safety (1SXXX)	.18	.19	.34
Scientific/Research (61XX)	.04	.27	.89
Security Forces (Military Police) (3PXXX)	.59	.17	.00
Security Forces (31XX)	.94	.21	.00
Services (34XX)	.24	.18	.20
Services (3MXXX)	.25	.15	.09
Space, Missile, and Command & Control (13XX)	.36	.15	.02
Special Investigations (OSI) (7SXXX)	.98	.18	.00
Supply (2SXXX)	.26	.15	.08
Support Commander (30XX)	.93	.24	.00
Surgery (45XX)	.55	.24	.02
TCN Escort	.14	.16	.37
Transportation & Vehicle Maintenance (2TXXX)	–.01	.14	.95
Unallotted Airman (9UXXX)	.31	.23	.17
Visual Information (3VXXX)	.35	.19	.07
Weather (15XX)	.13	.23	.55
Weather (1WXXX)	–.12	.16	.44

Table F.16
Regression Results for AFSC, Grade, Deployment Location, and Deployed Mission Predicting Importance Ratings for Changing Behavior to Fit Cultural Context—SHORT

Changing Behavior to Fit Cultural Context—SHORT	Unstandardized Coefficient	Standard Error	*p*
R-squared = .09			
Constant	2.54	.11	.00
Grade (1=low, 2=high)	–.07	.06	.20
Deployment location (1=IRAQ/AFG, 2=Other)	.20	.06	.00
Deployed mission (Flying Ops = comparison group)			
Base support	.17	.10	.07
CAOC or AOC	.03	.16	.85
Medical	.12	.09	.20
Other ops	.28	.10	.00
Staff above Wing level	.66	.11	.00
Unknown	.19	.22	.40
AFSC (Pilot = comparison group)			
Special Investigations (OSI) (AFSC 71XX)	1.35	.20	.00
Acquisition (63XX)	.70	.31	.02
Aerospace Medicine (48XX)	.44	.18	.02
Aircraft Maintenance (21XX)	.30	.14	.03
Aircrew Operations (1AXXX)	.26	.15	.08
Aircrew Protection (1TXXX)	.70	.13	.00
Biomedical Clinician (42XX)	.47	.19	.01
Biomedical Specialists (43XX)	.56	.22	.01
Cardiopulmonary Laboratory (4HXXX)	.83	.27	.00
Chaplain (52XX)	.74	.17	.00
Chaplain Assistant (5RXXX)	.49	.20	.02
Civil Engineer (32XX)	.62	.19	.00
Civil Engineering (3EXXX)	.36	.15	.02
Comm-Information Systems (33XX)	.23	.14	.11
Command & Control Systems Operations (1CXXX)	.31	.13	.02
Communications & Computer Systems (3CXXX)	.28	.15	.06
Communications & Electronics (2EXXX)	.27	.13	.05
Contracting (64XX)	1.08	.15	.00
Contracting (6CXXX)	.91	.15	.00
Finance (65XX)	.69	.21	.00
Financial (6FXXX)	.67	.15	.00
First Sergeant (8FXXX)	.50	.18	.01
Fuels (2FXXX)	.49	.15	.00
Group Superintendent (9GXXX)	.67	.24	.01
Health Services Administrator (41XX)	.47	.19	.01
Information Management (3AXXX)	.41	.13	.00
Intelligence (14XX)	.44	.14	.00

Table F.16—continued

Changing Behavior to Fit Cultural Context—SHORT	Unstandardized Coefficient	Standard Error	*p*
Intelligence (1NXXX)	.68	.13	.00
International Affairs/FAO (16XX)	.68	.18	.00
Law (51XX)	.38	.19	.05
Logistics Plans (2GXXX)	.13	.16	.39
Maintenance Management Systems (2RXXX)	.54	.16	.00
Manned Aerospace Maintenance (2AXXX)	.33	.15	.03
Manpower—Personnel (37XX)	.53	.17	.00
Medical (4AXXX)	.52	.14	.00
Medical Laboratory (4TXXX)	.48	.25	.06
Medical Service (4NXXX)	.46	.15	.00
Mental Health Services (4CXXX)	.22	.29	.45
Mission Support (3SXXX)	.43	.15	.01
Munitions & Weapons (2WXXX)	.23	.13	.08
Navigator (12XX)	.19	.14	.18
No AFSC Provided	.44	.21	.03
Nurse (46XX)	.60	.14	.00
Paralegal (5JXXX)	.53	.19	.01
Physician (44XX)	.60	.17	.00
Postal Specialist (8MXXX)	.55	.18	.00
Precision Measurement Equipment Lab (2PXXX)	.61	.19	.00
Public Affairs (35XX)	.91	.18	.00
Public Affairs (3NXXX)	.96	.16	.00
Safety (1SXXX)	.48	.19	.01
Scientific/Research (61XX)	.42	.21	.05
Security Forces (Military Police) (3PXXX)	.74	.17	.00
Security Forces (31XX)	1.14	.15	.00
Services (34XX)	.64	.16	.00
Services (3MXXX)	.56	.15	.00
Space, Missile, and Command & Control (13XX)	.49	.14	.00
Special Investigations (OSI) (7SXXX)	1.42	.18	.00
Supply (2SXXX)	.40	.15	.01
Support Commander (30XX)	1.04	.20	.00
Surgery (45XX)	.75	.23	.00
TCN Escort	.53	.19	.01
Transportation & Vehicle Maintenance (2TXXX)	.30	.15	.04
Unallotted Airman (9UXXX)	.57	.25	.02
Visual Information (3VXXX)	.63	.16	.00
Weather (15XX)	.48	.25	.05
Weather (1WXXX)	.11	.15	.46

Table F.17
Regression Results for AFSC, Grade, Deployment Location, and Deployed Mission Predicting Importance Ratings for Self-Initiated Learning—SHORT

Self-Initiated Learning—SHORT	Unstandardized Coefficient	Standard Error	*p*
R-squared = .13			
Constant	2.64	.10	.00
Grade (1=low, 2=high)	–.13	.06	.02
Deployment location (1=IRAQ/AFG, 2=Other)	.15	.05	.01
Deployed mission (Flying Ops = comparison group)			
Base support	.04	.09	.68
CAOC or AOC	–.22	.17	.22
Medical	.10	.09	.26
Other ops	.27	.10	.01
Staff above Wing level	.61	.10	.00
Unknown	.30	.22	.18
AFSC (Pilot = comparison group)			
Special Investigations (OSI) (AFSC 71XX)	1.18	.18	.00
Acquisition (63XX)	.67	.30	.03
Aerospace Medicine (48XX)	.43	.18	.02
Aircraft Maintenance (21XX)	.26	.13	.06
Aircrew Operations (1AXXX)	.12	.14	.39
Aircrew Protection (1TXXX)	.67	.13	.00
Biomedical Clinician (42XX)	.53	.19	.00
Biomedical Specialists (43XX)	.51	.18	.00
Cardiopulmonary Laboratory (4HXXX)	.72	.26	.01
Chaplain (52XX)	.84	.16	.00
Chaplain Assistant (5RXXX)	.59	.22	.01
Civil Engineer (32XX)	.62	.14	.00
Civil Engineering (3EXXX)	.34	.15	.02
Comm-Information Systems (33XX)	.31	.14	.03
Command & Control Systems Operations (1CXXX)	.28	.12	.02
Communications & Computer Systems (3CXXX)	.16	.15	.28
Communications & Electronics (2EXXX)	.08	.13	.55
Contracting (64XX)	1.16	.15	.00
Contracting (6CXXX)	.95	.14	.00
Finance (65XX)	.51	.26	.05
Financial (6FXXX)	.72	.15	.00
First Sergeant (8FXXX)	.35	.17	.04
Fuels (2FXXX)	.28	.14	.05
Group Superintendent (9GXXX)	.51	.23	.03
Health Services Administrator (41XX)	.55	.18	.00
Information Management (3AXXX)	.33	.13	.01
Intelligence (14XX)	.45	.14	.00

Table F.17—continued

Self-Initiated Learning—SHORT	Unstandardized Coefficient	Standard Error	*p*
Intelligence (1NXXX)	.62	.13	.00
International Affairs/FAO (16XX)	.79	.19	.00
Law (51XX)	.42	.19	.03
Logistics Plans (2GXXX)	.09	.16	.58
Maintenance Management Systems (2RXXX)	.33	.16	.05
Manned Aerospace Maintenance (2AXXX)	.13	.15	.36
Manpower—Personnel (37XX)	.54	.16	.00
Medical (4AXXX)	.57	.15	.00
Medical Laboratory (4TXXX)	.59	.25	.02
Medical Service (4NXXX)	.58	.14	.00
Mental Health Services (4CXXX)	.19	.23	.42
Mission Support (3SXXX)	.48	.15	.00
Munitions & Weapons (2WXXX)	.11	.13	.38
Navigator (12XX)	.17	.14	.22
No AFSC Provided	.60	.20	.00
Nurse (46XX)	.59	.13	.00
Paralegal (5JXXX)	.53	.17	.00
Physician (44XX)	.57	.19	.00
Postal Specialist (8MXXX)	.72	.15	.00
Precision Measurement Equipment Lab (2PXXX)	.33	.18	.06
Public Affairs (35XX)	.82	.20	.00
Public Affairs (3NXXX)	.95	.15	.00
Safety (1SXXX)	.39	.17	.03
Scientific/Research (61XX)	.57	.22	.01
Security Forces (Military Police) (3PXXX)	.82	.16	.00
Security Forces (31XX)	1.02	.15	.00
Services (34XX)	.75	.17	.00
Services (3MXXX)	.52	.14	.00
Space, Missile, and Command & Control (13XX)	.53	.14	.00
Special Investigations (OSI) (7SXXX)	1.30	.18	.00
Supply (2SXXX)	.34	.14	.02
Support Commander (30XX)	1.06	.22	.00
Surgery (45XX)	.94	.21	.00
TCN Escort	.46	.19	.02
Transportation & Vehicle Maintenance (2TXXX)	.15	.14	.28
Unallotted Airman (9UXXX)	.68	.25	.01
Visual Information (3VXXX)	.63	.18	.00
Weather (15XX)	.53	.20	.01
Weather (1WXXX)	.22	.15	.13

Table F.18
Regression Results for AFSC, Grade, Deployment Location, and Deployed Mission Predicting Importance Ratings for Applying Appropriate Social Etiquette—SHORT

Applying Appropriate Social Etiquette—SHORT	Unstandardized Coefficient	Standard Error	*p*
R-squared = .13			
Constant	2.74	.12	.00
Grade (1=low, 2=high)	–.08	.05	.10
Deployment location (1=IRAQ/AFG, 2=Other)	.27	.05	.00
Deployed mission (Flying Ops = comparison group)			
Base support	.14	.09	.12
CAOC or AOC	.05	.19	.78
Medical	.20	.09	.03
Other ops	.39	.09	.00
Staff above Wing level	.65	.10	.00
Unknown	.41	.22	.07
AFSC (Pilot = comparison group)			
Special Investigations (OSI) (AFSC 71XX)	1.25	.19	.00
Acquisition (63XX)	.61	.30	.04
Aerospace Medicine (48XX)	.51	.19	.01
Aircraft Maintenance (21XX)	.25	.15	.09
Aircrew Operations (1AXXX)	.22	.16	.16
Aircrew Protection (1TXXX)	.59	.15	.00
Biomedical Clinician (42XX)	.42	.20	.04
Biomedical Specialists (43XX)	.45	.20	.03
Cardiopulmonary Laboratory (4HXXX)	.78	.34	.02
Chaplain (52XX)	.83	.16	.00
Chaplain Assistant (5RXXX)	.53	.18	.00
Civil Engineer (32XX)	.61	.17	.00
Civil Engineering (3EXXX)	.26	.16	.11
Comm-Information Systems (33XX)	.25	.16	.13
Command & Control Systems Operations (1CXXX)	.27	.14	.05
Communications & Computer Systems (3CXXX)	.16	.16	.33
Communications & Electronics (2EXXX)	.15	.15	.32
Contracting (64XX)	1.23	.15	.00
Contracting (6CXXX)	1.04	.15	.00
Finance (65XX)	.67	.22	.00
Financial (6FXXX)	.71	.15	.00
First Sergeant (8FXXX)	.49	.19	.01
Fuels (2FXXX)	.39	.17	.02
Group Superintendent (9GXXX)	.59	.25	.02
Health Services Administrator (41XX)	.42	.19	.03
Information Management (3AXXX)	.38	.14	.01
Intelligence (14XX)	.38	.15	.01

Table F.18—continued

Applying Appropriate Social Etiquette—SHORT	Unstandardized Coefficient	Standard Error	*p*
Intelligence (1NXXX)	.59	.15	.00
International Affairs/FAO (16XX)	.65	.20	.00
Law (51XX)	.39	.19	.04
Logistics Plans (2GXXX)	.02	.17	.90
Maintenance Management Systems (2RXXX)	.48	.18	.01
Manned Aerospace Maintenance (2AXXX)	.11	.15	.45
Manpower—Personnel (37XX)	.60	.18	.00
Medical (4AXXX)	.45	.16	.00
Medical Laboratory (4TXXX)	.46	.29	.11
Medical Service (4NXXX)	.39	.15	.01
Mental Health Services (4CXXX)	.19	.27	.49
Mission Support (3SXXX)	.32	.16	.05
Munitions & Weapons (2WXXX)	.11	.14	.44
Navigator (12XX)	.14	.15	.37
No AFSC Provided	.49	.21	.02
Nurse (46XX)	.58	.15	.00
Paralegal (5JXXX)	.53	.19	.01
Physician (44XX)	.71	.18	.00
Postal Specialist (8MXXX)	.42	.16	.01
Precision Measurement Equipment Lab (2PXXX)	.47	.21	.03
Public Affairs (35XX)	.95	.19	.00
Public Affairs (3NXXX)	.94	.16	.00
Safety (1SXXX)	.41	.20	.04
Scientific/Research (61XX)	.48	.23	.04
Security Forces (Military Police) (3PXXX)	.73	.17	.00
Security Forces (31XX)	1.08	.16	.00
Services (34XX)	.91	.17	.00
Services (3MXXX)	.56	.15	.00
Space, Missile, and Command & Control (13XX)	.47	.15	.00
Special Investigations (OSI) (7SXXX)	1.40	.17	.00
Supply (2SXXX)	.30	.15	.05
Support Commander (30XX)	1.17	.22	.00
Surgery (45XX)	.85	.20	.00
TCN Escort	.55	.18	.00
Transportation & Vehicle Maintenance (2TXXX)	.23	.15	.13
Unallotted Airman (9UXXX)	.71	.26	.01
Visual Information (3VXXX)	.64	.18	.00
Weather (15XX)	.41	.24	.09
Weather (1WXXX)	.07	.17	.66

APPENDIX G

Open-Ended Comment Themes

Allied Operations: Remark mentioned working with foreign military.

Better Training—General Training: Remark stated need for better training but did not state specific content of training. Specific-content remarks are coded under appropriate headings (Allied Operations, Combat Skills, etc.)

Business Practices: Remark mentioned aspects of business. NOTE: The survey's initial 14 skills contained "negotiate effectively." Comments were coded as Business Practices if they mentioned aspects of business outside negotiation.

Combat Skills: Remark contained comment about training for convoy operations, firearms, escape and evasion, threat recognition, etc. These comments were distinguished from general Force Protection comments.

Educate/Train Foreign Military: Remark commented on training foreign military forces.

Force Protection—Awareness: Remark commented on need to be able to identify dangerous situations or need to avoid standing out as an American military member. These comments were not characterized by references to combat operations.

Joint Operations: Self-explanatory. Remark commented on working with Army, Navy, or Marine Corps personnel.

People Skills ("Common Sense"): Remark discussed relating to people across cultural barriers. Some respondents used the term *common sense*, so we included that phrase.

Role of Interpreter: Self-explanatory. Some respondents used the term *translator* instead of *interpreter*.

Role of Third-Country Nationals (TCNs): Remark commented on the role of third-country nationals working within U.S. military bases overseas. Remark usually correlated with those who commented on working inside the wire.

Transportation Skills (Non-Hostile Environment): Remark mentioned requirement for training on convoy operations, driving, or using public transportation.

Understand Culture: Remark singled out requirement to understand cultural component of deployment location and/or adversary.

Understand Religious Differences: Remark singled out requirement to understand religious component of deployment location and/or adversary.

Understand U.S. Policy/Mission: Remark stated requirement to understand and/or be able to explain the reason for U.S. presence in the deployed location.

Worked Inside the Wire: Comment stated that the respondent did not leave the base to do his/her part of the mission.

Miscellaneous: Comments did not fit into any of the identified themes and generally seemed to have little applicability to the survey.

References

Abbe, A., L. M. V. Gulick, and J. L. Herman (2007). *Cross-Cultural Competence in Army Leaders: A Conceptual and Empirical Foundation*, Study Report 2008-01, United States Army Research Institute for the Behavioral and Social Sciences.

Alliger, G. M., and E. A. Janak (1989). "Kirkpatrick's Levels of Training Criteria: Thirty Years Later," *Personnel Psychology*, 41, 331–342.

Alliger, G. M., S. I. Tannenbaum, W. Bennett, H. Traver, and A. Shotland (1997). "A Meta-Analysis of the Relations Among Training Criteria," *Personnel Psychology*, 50, 341–358.

Arthur, W., Jr., and W. Bennett, Jr. (1995). "The International Assignee: The Relative Importance to Factors Perceived to Contribute to Success," *Personnel Psychology*, 48, 99–114.

Bandura, A. (1977). *Social Learning Theory*, Englewood Cliffs, N.J.: Prentice Hall.

Bhaskar-Shrinivas, P., D. A. Harrison, M. A. Shaffer, and D. M. Luk (2005). "Input-Based and Time-Based Models of International Adjustment: Meta-Analytic Evidence and Theoretical Extension," *Academy of Management Journal*, 48, 257–281.

Bhawuk, D. P. S. (1998). "The Role of Culture Theory in Cross-Cultural Training: A Multimethod Study of Culture-Specific, Culture General, and Culture Theory-Based Assimilators," *Journal of Cross-Cultural Psychology*, 29, 630–655.

Birdseye, M. G., and J. S. Hill (1995). "Individual, Organizational/Work and Environmental Influences on Expatriate Turnover Tendencies: An Empirical Study," *Journal of International Business Studies*, 26, 787–813.

Black, J. S., and M. Mendenhall (1990). "Cross-Cultural Training Effectiveness: A Review and a Theoretical Framework for Future Research," *Academy of Management Review*, 15, 113–136.

Black, J. S., and L. W. Porter (1991). "Managerial Behaviors and Job Performance: A Successful Manager in Los Angeles May Not Succeed in Hong Kong," *Journal of International Business Studies*, 22, 99–113.

Black. J. S., M. Mendenhall, and G. Oddou (1991). "Toward a Comprehensive Model of International Adjustment: An Integration of Multiple Theoretical Perspectives," *The Academy of Management Review*, 16, 291–317.

Brady, R. A. (2006). "Presentation to the Military Personnel Subcommittee, Committee on Armed Services, United States House of Representatives; Subject: Recruiting and Retention and Military Personnel Policy, Benefits and Compensation Overview." Statement of: Lieutenant General Roger A. Brady, Deputy Chief of Staff, Manpower & Personnel, United States Air Force, April 6, 2006. As of December 31, 2008, available at: http://armedservices.house.gov/comdocs/schedules/4-6-06StatementBrady.pdf

Brett, J. M., and S. Okumura (1998). "Inter- and Intracultural Negotiation: U.S. and Japanese Negotiators," *Academy of Management Journal*, 41, 495–510.

Bush, G. W. (2006). "Remarks by President Bush to the U.S. University Presidents Summit on International Education," Washington, D.C.: U.S. Department of State. January 5, 2006. As of December 15, 2008: http://www.state.gov/r/summit/58734.htm

Caligiuri, P. M. (2000). "The Big Five Personality Characteristics as Predictors of Expatriate's Desire to Terminate the Assignment and Supervisor-Rated Performance," *Personnel Psychology*, 53, 67–88.

Campbell, J. P., and N. R. Kuncel (2001). "Individual and Team Training," in N. Anderson, D. S. Ones, H. K. Sinangil, and C. Viswesvaran (eds.), *Handbook of Industrial, Work, and Organizational Psychology*, 278–312.

Church, A. T. (1982). "Sojourner Adjustment," *Psychological Bulletin*, 91, 540–572.

Cochran, W. G. (1977). *Sampling Techniques*, 3rd ed., New York, N.Y.: John Wiley.

Cooper, C. L., P. J. Dewe, and M. O'Driscoll (2001). *Organizational Stress: A Review and Critique of Theory, Research, and Applications*, Thousand Oaks, Calif.: Sage.

Cross, S. E. (1995). "Self-Construal's, Coping, and Stress in Cross-Cultural Adaptation," *Journal of Cross-Cultural Psychology*, 26, 673–697.

Cushner, K., and R. W. Brislin (1996). *Intercultural Interactions: A Practical Guide* (2nd ed.), Thousand Oaks, Calif.: Sage.

Cushner, K., and D. Landis (1996). "The Intercultural Sensitizer," in D. Landis and R. S. Bhagat (eds.), *Handbook of Intercultural Training*, 2nd ed., Thousand Oaks, Calif.: Sage. 185–202.

Dinges, N. G., and K. D. Baldwin (1996). "Intercultural Competence: A Research Perspective," in D. Landis and R. S. Bhagat (eds.), *Handbook of Intercultural Training*, 2nd ed., Thousand Oaks, Calif.: Sage, 106–123.

DoD—*See* U.S. Department of Defense

Doney, P. M., J. P. Cannon, and M. R. Mullen (1998). "Understanding the Influence of National Culture on the Development of Trust," *The Academy of Management Review*, 23, 601–620.

Fiedler, F. E., T. Mitchell, and H. Triandis (1971). "The Culture Assimilator: An Approach to Cross-Cultural Training," *Journal of Applied Psychology*, 55, 95–102.

Gelfand, M. J., and N. Dyer (2000). "A Cultural Perspective on Negotiation: Progress, Pitfalls, and Prospects," *Applied Psychology: An International Review*, 49, 62–100.

Gelfand, M. J., M. Erez, and Z. Aycan (2007). "Cross-Cultural Organizational Behavior," *Annual Review of Psychology*, 58, 479-514.

Glaser, R. (1984). "Education and Thinking: The Role of Knowledge," *American Psychologist*, 39, 93–104.

Goldstein, I. L. (1991). "Training in Work Organizations," in M. D. Dunnette and L. M. Hough (eds.), *Handbook of Industrial and Organizational Psychology*, 2nd ed., Palo Alto, Calif.: Consulting Psychologists Press, 106–123.

Greeno, J. G., A. M. Collins, and L. B. Resnick (1996). "Cognition and Learning," in D. C. Berliner and R. C. Calfee (eds.), *Handbook of Educational Psychology*, New York, N.Y.: Simon & Schuster Macmillan, 15–46.

Gudykunst, W. B., R. M. Guzley, and M. R. Hammer (1996). "Measuring Impacts of Cross-Cultural Training," in D. Landis and R. S. Bhagat (eds.), *Handbook of Intercultural Training*, 2nd ed., Thousand Oaks, Calif.: Sage, 165–182.

Hammer, M. R., M. J. Bennett, and R. Wiseman (2003). "Measuring Intercultural Sensitivity: The Intercultural Development Inventory," *International Journal of Intercultural Relations*, 27, 421–443.

Harrison, J. K. (1992). "Individual and Combined Effects of Behavior Modeling and the Cultural Assimilator in Cross-Cultural Management Training," *Journal of Applied Psychology*, 77, 952–962.

Hofstede, G., and G. J. Hofstede (2005). *Cultures and Organizations: Software of the Mind*. New York, N.Y.: McGraw-Hill.

Hom, P. W., and R. W. Griffeth (1995). *Employee Turnover*. Cincinnati, Ohio: South-Western College Publishing.

Janssens, M. (1995). "Intercultural Interaction: A Burden of International Managers," *Journal of Organizational Behavior*, 16, 155–167.

Joint Chiefs of Staff (2005). *Officer Professional Military Education Policy (OPMEP)*, CJCSI 1800.01C, December 22, 2005 (current as of August 7, 2007). As of December 9, 2008, available through: http://www.dtic.mil/cjcs_directives/cjcs/instructions.htm

Kraiger, K. (2003). "Perspectives in Training and Development," in W. C. Borman, D. R. Ilgen, and R. J. Klimoski (eds.), *Handbook of Psychology*, 12, 171–192.

Lorenz, S. R. (2007). "Transforming Air Force Education for the Long War and Beyond," *Air & Space Power Journal*, Summer 2007. As of December 15, 2008:
http://www.airpower.maxwell.af.mil/airchronicles/apj/apj07/sum07/lorenz.html

Mak, A. S., M. J. Westwood, F. I. Ishiyama, and M. C. Barker (1999). "Optimizing Conditions for Learning Sociocultural Competencies for Success," *International Journal of Intercultural Relations*, 23, 77–90.

Manacapilli, T., C. M. Hardison, B. Gifford, A. Bailey, and A. Bower (2007). *Common Battlefield Training for Airmen*, MG-624-AF, Santa Monica, Calif.: RAND Corporation. As of December 15, 2008:
http://www.rand.org/pubs/monographs/MG624/

Mendenhall, M. E., G. K. Stahl, I. Ehnert, G. Oddou, J. S. Osland, and T. M. Kuhlmann (2004). "Evaluation Studies of Cross-Cultural Training Programs: A Review of the Literature from 1988 to 2000," in D. Landis, J. M. Bennett, and M. J. Bennett (eds.), *Handbook of Intercultural Training*, 3rd ed., Thousand Oaks, Calif.: Sage, 129–143.

Mol, S. T., M. Ph. Born, M. E. Willemsen, and H. T. Van Der Molen (2005). "Predicting Expatriate Job Performance for Selection Purposes: A Quantitative Review," *Journal of Cross-Cultural Psychology*, 36, 590–620.

Molinsky, A. (2005). "Language Fluency and the Evaluation of Cultural Faux Pas: Russians Interviewing for Jobs in the United States," *Social Psychology Quarterly*, 68, 103–120.

Morris, M. A., and C. Robie (2001). "A Meta-Analysis of the Effects of Cross-Cultural Training on Expatriate Performance and Adjustment," *International Journal of Training and Development*, 5, 112–125.

Morris, M. W., K. Y. Williams, K. Leung, R. Larrick, M. T. Mendoza, D. Bhatnagar, J. Li, M. Kondo, J. Luo, and J. Hu. (1998). "Conflict Management Style: Accounting for Cross-National Differences," *Journal of International Business Studies*, 29, 729–747.

Nunnally, J. C., and I. H. Bernstein (1994). *Psychometric Theory*, 3rd ed., New York, N.Y.: McGraw-Hill.

Ohbuchi, K., and Y. Takahashi (1994). "Cultural Styles of Conflict Management in Japanese and Americans: Passivity, Covertness, and Effectiveness of Strategies," *Journal of Applied Social Psychology*, 24, 1345–1366.

Ones, D. S., and C. Viswesvaran (1997). "Personality Determinants in the Prediction of Aspects of Expatriate Job Success," *New Approaches to Employee Management*, 4, 63–92.

Rice, C. R. (2006). "Remarks at the Summit of U.S. University Presidents on International Education," Washington, D.C.: U.S. Department of State, January 5, 2006. As of December 15, 2008:
http://www.state.gov/secretary/rm/2006/58735.htm

Shaffer, M. A., D. A. Harrison, H. Gregersen, J. S. Black, and L. A. Ferzandi (2006). "You Can Take It With You: Individual Differences and Expatriate Effectiveness," *Journal of Applied Psychology*, 91, 109–125.

Sheer, V. C., and L. Chen (2003). "Successful Sino-Western Business Negotiation: Participant's Accounts of National and Professional Cultures," *Journal of Business Communication*, 40, 50–85.

Sullivan, J., R. B. Peterson, N. Kameda, and J. Shimada (1981). "The Relationship Between Conflict Resolution Approaches and Trust—A Cross Cultural Study," *The Academy of Management Journal*, 24, 803–815.

Ting-Toomey, S. (2004). "Translating Conflict Face-Negotiation Theory into Practice," in D. Landis, J. M. Bennett, and M. J. Bennett (eds.), *Handbook of Intercultural Training* (3rd ed.), Thousand Oaks, Calif.: Sage, 217–248.

Triandis, H. C. (1996). "The Psychological Measurement of Cultural Syndromes," *American Psychologist*, 51, 407–415.

Triandis, H. C., and D. P. S. Bhawuk (1997). "Culture Theory and the Meaning of Relatedness," in P. C. Earley and M. Erez (eds.), *New Perspectives on International Industrial/Organizational Psychology*, San Francisco, Calif.: Jossey-Bass, 13–52.

Tyler, T. R., E. A. Lind, and Y. J. Huo (2000). "Cultural Values and Authority Relations: The Psychology of Conflict Resolution Across Cultures," *Psychology, Public Policy, and Law*, 6, 1138–1163.

U.S. Department of Defense (2005). "Defense Language Transformation Roadmap," Washington, D.C., January 2005. As of December 15, 2008, available at: http://www.defenselink.mil/news/Mar2005/d20050330roadmap.pdf

U.S. Department of Defense (2006). "Quadrennial Defense Review Report," Washington, D.C., February 6, 2006. As of December 15, 2008, available at: http://www.defenselink.mil/qdr/report/Report20060203.pdf

U.S. Department of Defense (2007). "DoD Regional and Cultural Expertise Guidelines," Draft document provided to Air Force Directorate of Airman Development.

U.S. Department of State (2006). "National Security Language Initiative," Fact Sheet, Office of the Spokesman, Washington, D.C.: U.S. Department of State, January 5, 2006. As of December 15, 2008: http://www.state.gov/r/pa/prs/ps/2006/58733.htm

Ward, C., and A. Kennedy (1993). "Where's the 'Culture' in Cross-Cultural Transition? Comparative Studies of Sojourner Adjustment," *Journal of Cross-Cultural Psychology*, 24, 221–249.

Ward, C., Y. Okura, A. Kennedy, and T. Kojima (1998). "The U-Curve on Trial: A Longitudinal Study of Psychological and Sociocultural Adjustment During Cross-Cultural Transition," *International Journal of Intercultural Relations*, 22, 277–291.

White, M. J., and Y. Li (1991). "Second-Language Fluency and Person Perception in China and the United States," *Journal of Language and Social Psychology*, 10, 99–113.

Wible, D. S., and C. H. Hui (1985). "Perceived Language Proficiency and Person Perception," *Journal of Cross-Cultural Psychology*, 16, 206–222.

Williams, K. M., and J. L. Crafts (1997). "Inductive Job Analysis: The Job/Task Inventory Method," in D. L. Whetzel and G. R. Wheaton (eds.), *Applied Measurement Methods in Industrial Psychology*, Palo Alto, Calif.: Davies-Black Publishing.

Wolfowitz, P. (2003). Interview by Doug Jehl, *New York Times*. News transcript, U.S. Department of Defense, Office of the Assistant Secretary of Defense (Public Affairs), May 3, 2003. As of December 15, 2008: http://www.defenselink.mil/transcripts/transcript.aspx?transcriptid=2567

Worchel, S., and T. R. Mitchell (1972). "An Evaluation of the Effectiveness of the Culture Assimilator in Thailand and Greece," *Journal of Applied Psychology*, 56, 472–479.

Wynne, M. W. (2006). "Education and the Airman," *Letter to Airmen*, Michael W. Wynne, Secretary of the Air Force, April 12, 2006. As of December 15, 2008, available at: integrator.hanscom.af.mil/2006/April/04202006/LtA_Education.pdf